Operational amplifier experimental manual

G. B. Clayton
B.Sc., F.Inst.P.
Principal Lecturer in Physics
Liverpool Polytechnic, England

Operational amplifier experimental manual

Butterworths

London · Boston · Durban · Singapore · Sydney · Toronto · Wellington

First published 1983

British Library Cataloguing in Publication Data
Clayton, G.B.
Operational amplifier experimental manual.
1. Operational amplifiers
I. Title
621.3815'35 TK7871.58.06

ISBN 0-408-01240-4
ISBN 0-408-01239-0 Pbk

Filmset in Monophoto Times by
Northumberland Press Ltd. Gateshead
Printed in Great Britain by The Camelot Press Ltd, Southampton

Contents

Introduction

Operational amplifier experiments and the experimental circuit board

This manual describes experimental procedures for evaluating the many practical applications of integrated circuit operational amplifiers. All the experiments are based upon the use of a single test board, details of which are given. The board is designed so that test circuits can be rapidly connected using a system of jump leads terminated by sockets which snap onto pins mounted on the board. Diagrams showing the lead connections required for each experiment are given and test measurements which provide a means of evaluating circuit performance are suggested.

The manual is intended as a companion to the book *Operational Amplifiers Second Edition**. The experimental work which is suggested is best regarded as a reinforcement of the learning process. Experimentation should ideally be performed concurrently with a reading of both this manual and the companion text. If an increased understanding of operational amplifier applications is sought, experimentation should be performed 'knowingly'. If in an experiment a particular reading or waveform is observed, the experimenter should strive to know why he observes what he does observe in terms of the explanations, discussions and theoretical background which are provided briefly in this manual and in more detail in the companion text.

Before commencing experimental work the reader is advised to familiarize himself with the physical layout of the experimental circuit board. A diagram showing details of the board is given in Figure 1.

---

*Clayton, G. B., *Operational Amplifiers Second Edition*, Butterworth Scientific, London, 1979.

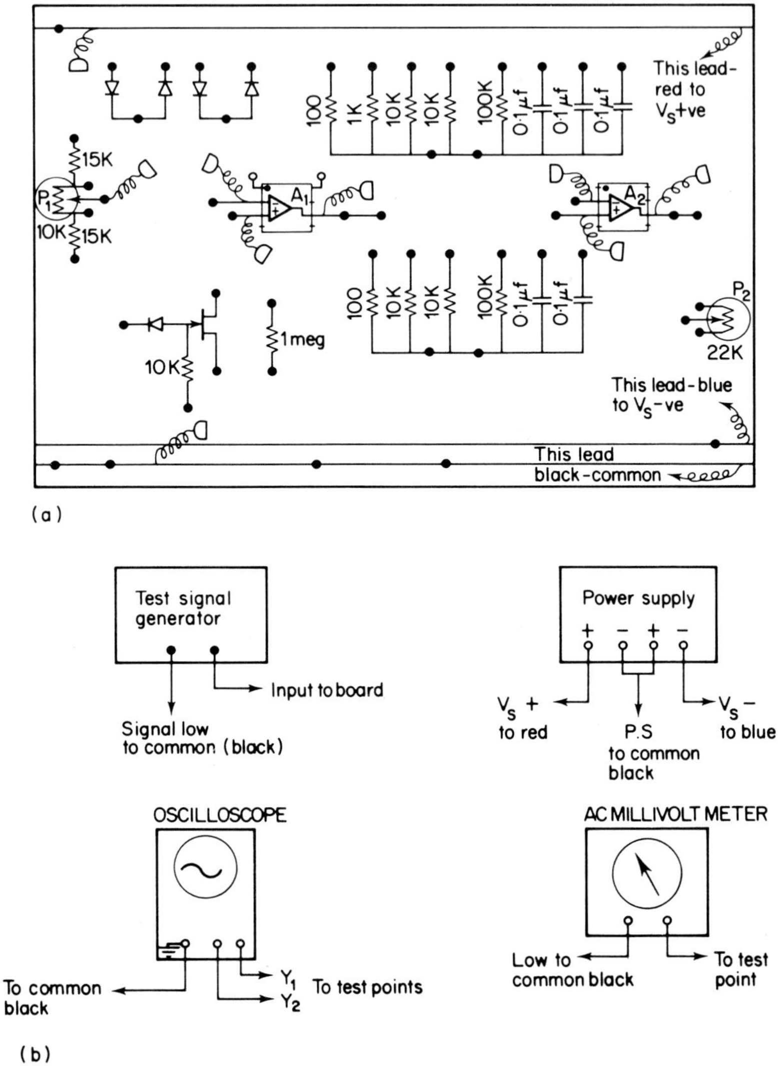

Figure 1 The experimental board: (a) location of components on board; (b) instrument connections to board

This should be checked against the actual board in order to find the locations of the various components. Provision for two operational amplifiers is made by two sockets which are mounted on the board. The standard integrated circuit operational amplifier pin configuration, is assumed: pin 1, offset trim and frequency compensation; pin 2, inverting input terminal; pin 3, non-inverting input terminal; pin 4, power supply negative; pin 5, offset trim; pin 6, output; pin 7, power supply positive; pin 8, frequency compensation or no connection.

From the point of view of experimental evaluation the choice of a particular amplifier type for use with the board is somewhat arbitrary. The amplifier type used will be found to influence particular aspects of circuit performance. These aspects will emerge as the student progresses with his reading and experimental work. Unless otherwise stated, it will be assumed that an internally compensated BiFet operational amplifier (e.g. type TL081 or similar) is to be used.

The board requires externally connected twin power supplies. Incorrect power supply connection causes damage to operational amplifiers; it is therefore important not to make mistakes in the power lead connections. The red lead should be connected to V_{s+ve} (+15 V), the black lead to power supply common (the line connecting the positive terminal of one supply to the negative terminal of the other), and the blue lead to V_{s-ve} (−15 V).

Input and output overload protection circuitry is incorporated in most modern operational amplifier devices. The protective circuitry is only operative when the amplifier is connected to its power supplies. Power supplies to the board should therefore always be switched on *before* applying test signals. Test signals are measured with respect to the power supply common line.

The main items of external test equipment required for use with the board are a variable frequency sine/square signal generator, a dual-trace oscilloscope, a high-impedance a.c. voltmeter (with a decibel (dB) scale) and a general purpose amps/volts/ohms meter.

The measurements suggested in the various experiments do not in any way represent an exhaustive investigation of circuit behaviour. The experimenter should pursue any further investigations which might occur to him. Once a familiarity with the experimental board layout has been achieved, it is a simple matter to change from one circuit configuration to another.

Readers wishing to perform the experiments without using the recommended experimental board can of course do so. If the recommended board is not available experimental circuits can be connected up using some alternative bread-boarding system. In this case the link connections which are shown in diagrams will not be directly applicable and circuits will need to be assembled using the given theoretical circuit diagrams as a guide.

The recommended board has the advantage of providing, in a permanent form, all required components thus avoiding time wastage in searching out component values. The method of interconnecting components which is used by the recommended board ensures good electrical contacts. It overcomes the problems associated with loose connections and poor contacts that can often be so troublesome in non-soldered circuits.

The recommended experimental board has been manufactured commercially and is available from:

E & L INSTRUMENTS LTD
Whitegate Industrial Estate
Wrexham LL13 8UG.

As a further aid in the learning process there are a series of C.A.L. programs available; they are intended for use by readers with access to a personal computer. The programs deal with the various aspects of operational amplifier behaviour covered in this manual and in its companion text (see Appendix). The programs are available from:

S.C.I.C.A.L. Software,
13 Kingsmead Road North,
Oxton,
Birkenhead,
Merseyside L43 6TB.

Experiment 1

Basic resistive feedback amplifier configurations

This experiment is intended to provide the student with an opportunity to achieve a practical familiarity with the basic operational amplifier feedback configurations. Factors which influence the finer aspects of circuit behaviour such as bandwidth limitations and other performance errors are examined in later experiments. Before performing Experiment 1 the student should read Chapter 1 of the companion text.

The unity-gain voltage follower

This is the simplest feedback configuration. The whole of the amplifier output signal is returned to its inverting input terminal. The amplifier output is forced to follow any signal voltage variation applied to its non-inverting input terminal.

Make the link connection shown in Figure 2a. Apply an input test signal (say a low-frequency sinusoidal signal with amplitude less than 10 V) and monitor the input and output signals with a dual-trace oscilloscope. Verify that the output and input signals are in phase and that the ratio (output signal amplitude)/(input signal amplitude) is unity.

The very high effective input impedance of the follower circuit can be verified by connecting a 1 MΩ resistor in series with the input test signal (Figure 2b). A negligible reduction in output signal amplitude signifies an input resistance much greater than 1 MΩ. Now increase the frequency of the input test signal (set the amplitude of

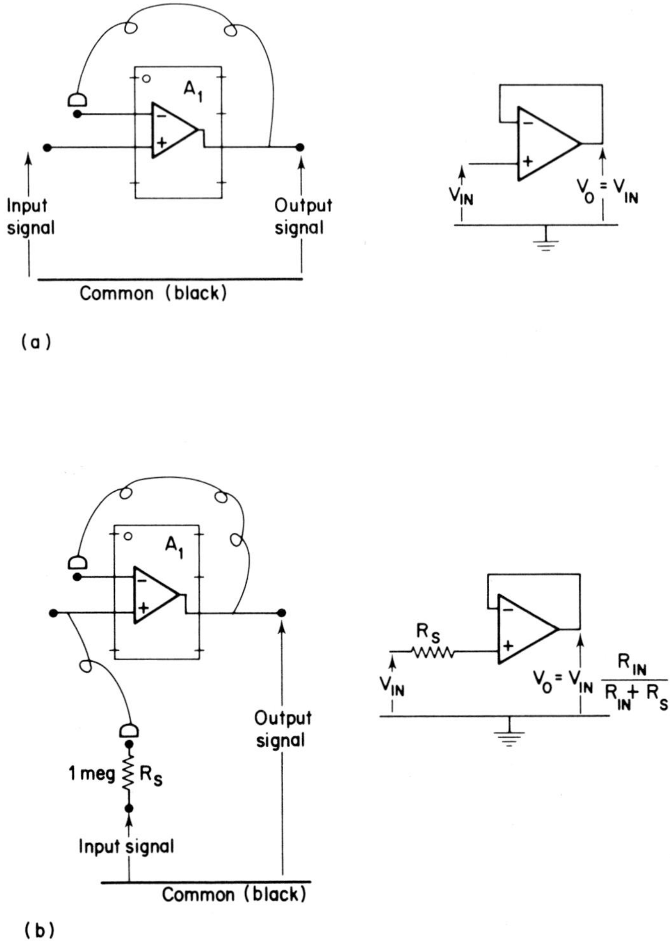

Figure 2 The unity-gain voltage follower

the signal to about 100 mV) and note the frequency at which the output amplitude drops by 3 dB (by 0.71). The input capacitance of the circuit forms a low-pass filter with the 1 MΩ resistor. The effective input capacitance may be calculated from

$$C_{\text{eff}} = 1/2\pi \times 10^6 f_{3\text{dB}}$$

The follower with gain

Make the link connections shown in Figure 3. Apply a low-frequency input test signal; monitor and measure the input and output signals with a dual-trace oscilloscope.

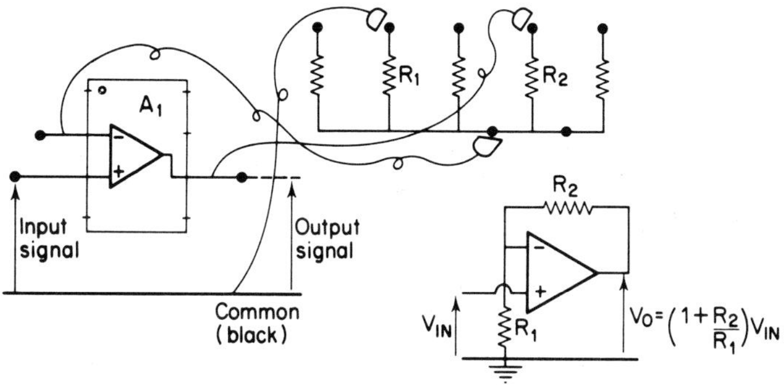

Figure 3 The follower with gain

Verify that the voltage gain of the circuit is

$$\frac{\text{output signal amplitude}}{\text{input signal amplitude}} = 1 + \frac{R_2}{R_1}$$

Several gain settings are available, obtained by making link connections to select resistors from the upper resistor bank on the board. For example,

$R_1 = R_2 = 10\ \text{k}\Omega$ makes gain = 2
$R_2 = 100\ \text{k}\Omega$, $R_1 = 10\ \text{k}\Omega$ makes gain = 11
$R_2 = 100\ \text{k}\Omega$, $R_1 = 1\ \text{k}\Omega$ makes gain = 101

The inverter

Make the link connections shown in Figure 4. Apply a low-frequency (say 200 Hz) input test signal; monitor and measure the input and output signals with a dual-trace oscilloscope.

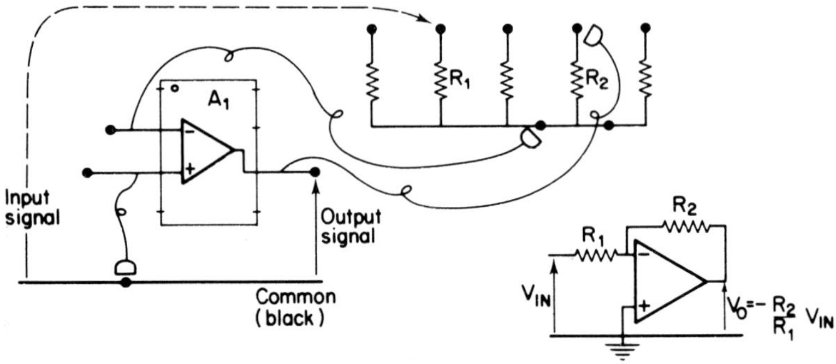

Figure 4 The inverter

Note that the output signal is 180° out of phase with the input signal. Verify that the voltage gain

$$\frac{\text{output signal amplitude}}{\text{input signal amplitude}} = -\frac{R_2}{R_1}$$

Try the several gain settings which are available by selecting resistors R_1 and R_2 from the upper bank of resistors. The inverting circuit can be given a gain less than unity.

The inverting adder

Make the link connections shown in Figure 5. Verify that the

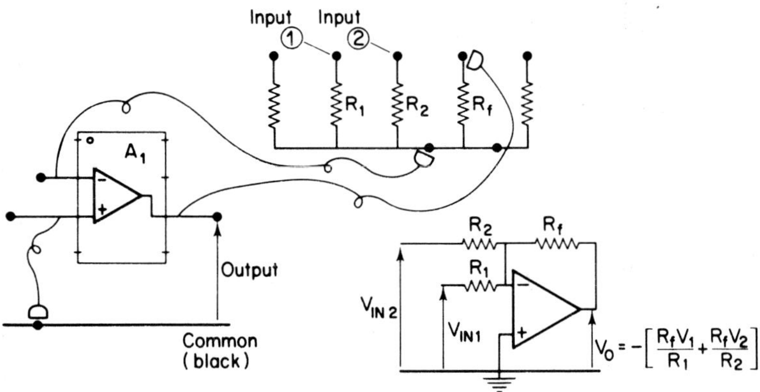

Figure 5 The inverting adder

output signal is determined by the sum of the two input signals V_1 and V_2 scaled according to the values selected for resistors R_1 and R_2. One of the input signals can be a low-frequency sine wave and the other can be a d.c. voltage derived from, say, potentiometer P_1 connected between the supply rails.

Exercises 1

1 Sketch circuit arrangements with component values for the following:
(a) a circuit having a voltage gain -100 and input resistance 1 kΩ;
(b) a circuit with its gain varied by a potentiometer setting in the range $+1$ to $+23$ (use a potentiometer value 22 kΩ);
(c) a circuit with its output voltage V_o related to the values of two input signals V_1 and V_2 by the equation $V_o = -(10V_1 + 0.1V_2)$.
Connect the circuits on the experimental board and verify their performance relationships.

Experiment 2

Offset errors. Measurement of amplifier input offset voltage and bias current

Before performing this experiment the student is advised to read Sections 2.10 and 3.1 of the companion text.

In the previous experiments with the follower and inverter, at the higher gain settings, it may have been observed that the amplifier output voltage was not exactly zero when the input signals were zero. A non-zero output with zero input represents an offset error. Offset errors are due to the combined effects of the input offset voltage and bias current of an operational amplifier. In this experiment measured values of the output offset error are used to deduce the values of the input offset voltage and bias current of the amplifier. The amplifier is connected in a feedback configuration with a high closed-loop gain. It is suggested that a bipolar input operational amplifier (e.g. type 741 or similar) should be used initially in this experiment.

Measurement of amplifier input offset voltage V_{io}

Make the link connections shown in Figure 6a. Measure the offset error at the amplifier output terminal. Because of the low value of the source resistance at the amplifier input terminals the output offset in the circuit is due almost entirely to the input offset voltage V_{io} of the operational amplifier.

The output voltage error

$$V_{o_1} = V_{io}\left(1 + \frac{R_2}{R_1}\right) = 1001\, V_{io}$$

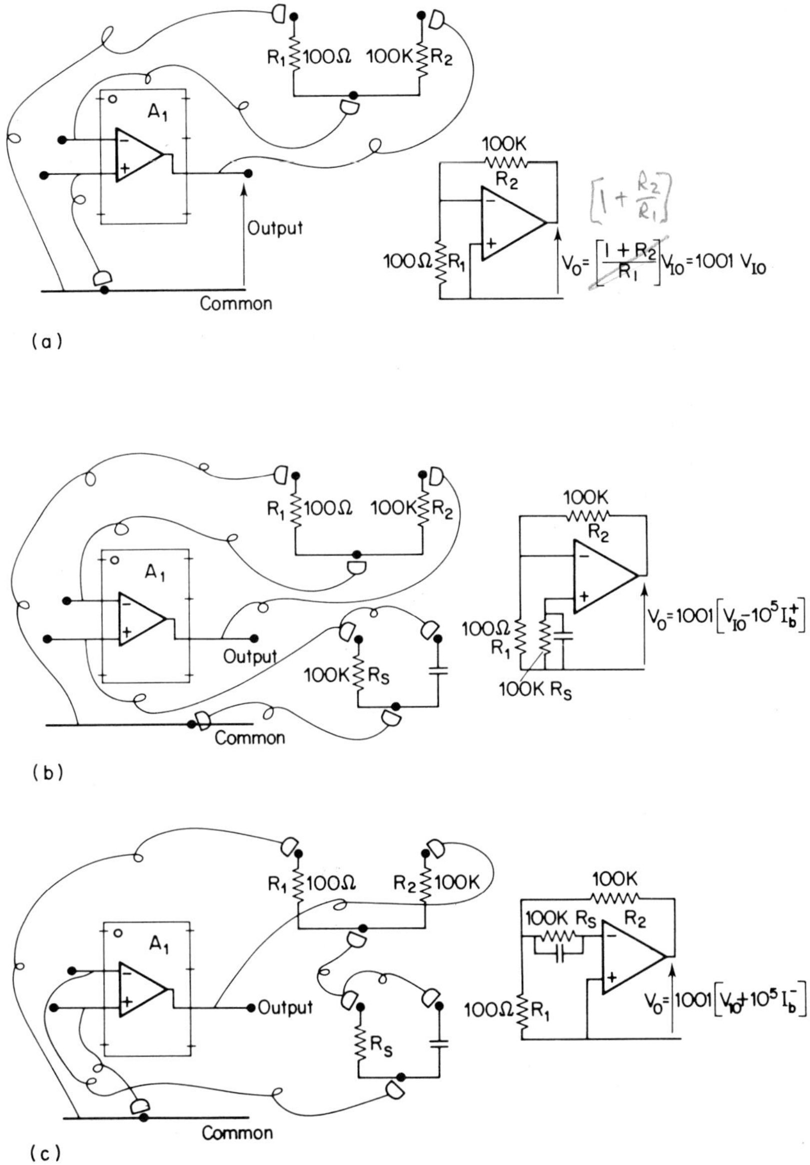

Figure 6 Measurement of (a) input offset voltage V_{io}, (b) bias current I_b^+ and (c) bias current I_b^-

Hence $V_{io} \approx V_{o_1}/1000$. For example, a measured output offset error of say 3 V signifies that the input offset voltage V_{io} of the amplifier is 3 mV. The offset may be of either polarity.

Measurement of the bias current drawn by the non-inverting input terminal, I_b^+

Make the link connections shown in Figure 6b. Measure the offset error at the amplifier output terminal. In this circuit the output offset is due to V_{io} and the voltage $-I_b^+ R_s^+$ which arises because of the bias current I_b^+ flowing through the resistor R_s^+ (100 kΩ) and which is applied to the non-inverting input.

The output offset error

$$V_{o_2} = (V_{io} - I_b^+ R_s^+)(1 + R_2/R_1)$$

Hence $I_b^+ \approx (V_{o_1} - V_{o_2})/10^3 \times 10^5$.

Measurement of the bias current drawn by the inverting input terminal, I_b^-

Make the link connections shown in Figure 6c. Measure the offset error at the amplifier output. In this circuit the output offset is due to the combined effects of V_{io} and $-I_b^- R_s^-$. The voltage $-I_b^- R_s^-$ is applied to the inverting input as a result of the bias current I_b^- flowing through the resistance R_s^- (100 kΩ).

$$V_{o_3} = (V_{io} + I_b^- R_s^-)(1 + R_2/R_1)$$

Hence $I_b^- \approx (V_{o_3} - V_{o_1})/10^3 \times 10^5$.

Note. Amplifier bias current as determined by a data sheet value is $I_b = (I_b^+ + I_b^-)/2$. Amplifier input offset current I_{io} is the difference between the bias currents drawn at the two inputs: $I_{io} = I_b^+ - I_b^-$.

Further measurements

Amplifier input offset voltage V_{io} and bias current I_b change with temperature and power supply voltage. They also exhibit a long-term variation with time. The dependence of the parameters on power supply voltages can be exhibited by repeating the above readings using different values of the power supply voltage to the board. The effect of changes in V_{s+ve} and V_{s-ve} should be found separately.

There is a considerable spread in the V_{io} and I_b values for different specimens of the same amplifier type. Plug in different 741's and find their V_{io} and I_b values.

Field effect transistor (FET) input operational amplifiers have a much smaller bias current than bipolar input amplifiers. Repeat the above experiment using a BiFet amplifier (e.g. TL081). You will probably find that introducing the 100 kΩ resistor in series with the input leads to the BiFet amplifier has a negligible effect on the observed output offset error. The test board, as it stands, is not suitable for the measurement of the very small bias currents which are drawn by FET input type operational amplifiers. (See Figure 28, Experiment 9, for an alternative method of measuring the bias current of an FET input operational amplifier.)

Exercises 2

1 If the values $V_{o_1} = -2.5$ V, $V_{o_2} = -8.5$ V, $V_{o_3} = +4.5$ V are obtained in this experiment, what are the values of the input offset voltage and bias current of the amplifier?

2 An operational amplifier has an input offset voltage of 2 mV and a bias current of 60 nA. What is the worst-case output offset error if the amplifier is used as:
(a) an inverter with input resistor 10 kΩ and feedback resistor 100 kΩ;
(b) an inverting adder with input resistors 10 kΩ and 100 kΩ and feedback resistor 100 kΩ?

Experiment 3

Measurement of the open-loop gain of an operational amplifier. Measurement of the frequency dependence of the open-loop gain of an internally frequency-compensated operational amplifier

An acquaintance with the material contained in Chapter 2, pages 16–38, and Section 3.5 of the companion text will be found advantageous for an understanding of this and the following experiment.

The open-loop gain of an operational amplifier is defined by the ratio (change of output voltage)/(change of input voltage), the input voltage being that measured directly between the inverting and non-inverting input terminals of the amplifier. Open-loop gain is normally specified for very slowly varying input signals and with the amplifier output supplying a specified load.

The open-loop gain of most operational amplifiers is very large (of the order 10^5) for low-frequency signals, but its value falls as the signal frequency is increased (roll-off). The frequency at which the open-loop gain of an amplifier has fallen to unity is called the unity-gain frequency (or the unity-gain bandwidth) of the amplifier. In the case of internally frequency-compensated amplifiers, open-loop gain roll-off normally follows a 20 dB/decade slope down to unity gain. The above characteristics of the open-loop gain of an operational amplifier are demonstrated in this experiment.

Make the link connections shown in Figure 7. Apply a low-frequency (<10 Hz) sinusoidal test signal at the point A. Monitor and measure the signals at points X and Y with a dual-trace oscilloscope. Deduce the value of the open-loop gain of the amplifier from

$$A_{OL} = \frac{\text{peak-to-peak signal at X}}{\text{peak-to-peak signal at Y}} \times 100$$

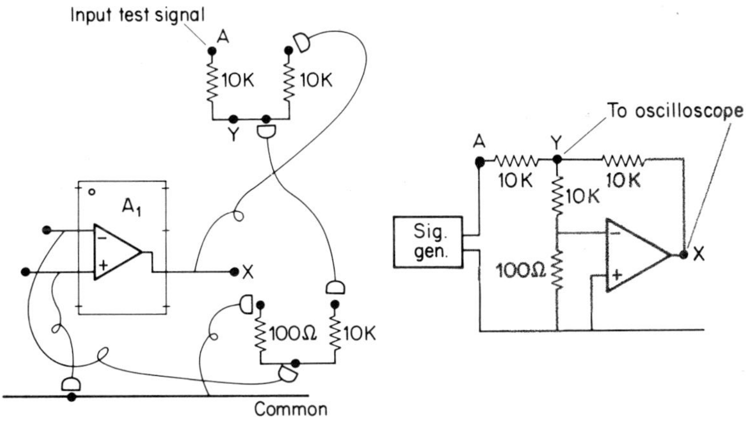

Figure 7 Measurement of open-loop gain

Note that the multiplying factor 100 arises because the signal which is observed at point Y is 100 times bigger than the signal which is actually applied between the differential input terminals of the amplifier.

Increase the frequency of the test signal and repeat the above measurements. A decrease in the open-loop gain of the amplifier will show itself as an increase in the signal amplitude observed at point Y. The signal observed at point X will be found to lag in phase on the signal at Y. At the higher frequencies, when the open-loop gain has fallen to say 100, the resistive potential divider at the amplifier input should be removed. The inverting input terminal of the amplifier should be linked directly to the point Y. The open-loop gain is then determined as the ratio of the signal amplitudes at X and Y.

A number of measurements should be made to allow a graph of the type shown in Figure 8 to be plotted. A five-decade log–linear piece of graph paper is needed. Gain measurements taken say at decade frequency intervals should be sufficient to show the general form of the gain roll-off. Voltage gain, expressed in dB, is plotted against logarithm of frequency:

$$A_{OL} \text{ (in dB)} = 20\log_{10}A_{OL}$$

If the results obtained in this experiment are to be a valid representation of the small-signal frequency response of the amplifier the

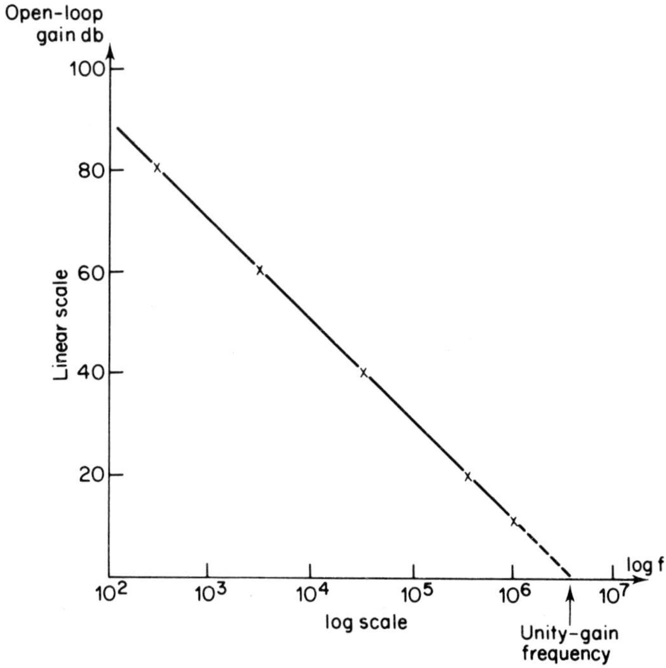

Figure 8 Open-loop gain roll-off plotted on log–linear graph paper

experimenter must ensure that no slew rate limiting of the amplifier output signal occurs. The onset of slew rate limiting is evidenced by distortion of the amplifier output signal (see Experiment 5). It is simply necessary to reduce the amplitude of the test signal in order to remove any slew rate distortion. To avoid slew rate limiting the amplitude of the amplifier output signal must be such that

$$\text{Peak-to-peak signal amplitude at X} < \frac{\text{slew rate}}{\pi f} \times 10^6 \text{ V}$$

The slew rate of an amplifier is normally expressed in V/μs. A 741 type amplifier has a slew rate of typically 0.5 V/μs. A TL081 BiFet amplifier has a typical slew rate of 13 V/μs. Thus for a 741 type amplifier, at a test frequency 10^6 Hz, to avoid slew rate limiting, the amplitude of a sinusoidal output signal must be less than $0.5 \times 10^6/\pi \times 10^6 = 160$ mV peak-to-peak. At a test frequency 10^5 Hz the peak-to-peak output must be less than 1.6 V.

Additional measurements

1. From observations of the waveforms at X and Y measure the phase shift in the amplifier open-loop gain as a function of frequency (see Figure 61, Experiment 20, for a method of measuring phase shift).
2. Find the effect on the amplifier open-loop gain of connecting different load resistor values between the amplifier output terminal and the zero voltage power supply common line (use values in the range say 1–10 kΩ).
3. Find the dynamic output voltage range of the amplifier by increasing the amplitude of the input signal until clipping of the output signal is observed. Observe the effect of the power supply values on the amplifier output voltage limits.

Exercises 3

1 An internally frequency-compensated operational amplifier has an open-loop gain at zero frequency of 10^5 and a unity-gain frequency of 3×10^6 Hz. Sketch Bode magnitude and phase approximations showing the frequency dependence of the open-loop gain.

2 An internally frequency-compensated operational amplifier has an open-loop gain of 2×10^5 at zero frequency and a unity-gain frequency of 2.5×10^6 Hz.
(a) What is the value of the open-loop gain of the amplifier at frequencies 2.5×10^5 Hz, 10^5 Hz, 10^4 Hz?
(b) At what frequency does the open-loop gain have a phase shift of 45° associated with it?

Experiment 4

Measurement of the small-signal dynamic response of an internally frequency-compensated operational amplifier, closed-loop bandwidth and rise time

Relevant sections of the companion text are Sections 2.4, 2.5, 2.6, 2.8 and 3.8.

An internally frequency-compensated operational amplifier when used with purely resistive feedback components has a small-signal closed-loop bandwidth which is determined by the product of the feedback fraction β and the unity-gain frequency f_1 of the amplifier:

$$\text{Closed-loop bandwidth } f_1^1 = \beta f_1 \tag{4.1}$$

The closed-loop bandwidth is the frequency at which the gain of the feedback amplifier is 3 dB down on its value at zero frequency.

At frequencies above the closed-loop bandwidth the roll-off in the closed-loop gain follows the roll-off in the open-loop gain.

The small-signal output rise time of an internally frequency-compensated amplifier, when used with a resistive feedback fraction β, is determined by the relationship

$$T_r \approx 1/3f_1^1 \text{ or } T_r \approx 1/3\beta f_1 \tag{4.2}$$

The above characteristics of the closed-loop behaviour of an operational amplifier are demonstrated by this experiment.

Make the link connections shown in Figure 9. The circuit is simply the inverting feedback configuration previously investigated in Experiment 1. Different values of β are established by selecting different values for R_2 and R_1 from the upper bank of resistors on the experimental board. For example, $R_2 = 100$ kΩ, $R_1 = 1$ kΩ, makes the signal gain $R_2/R_1 = -100$ and the feedback fraction $\beta = R_1/(R_1 + R_2) = 1/101$.

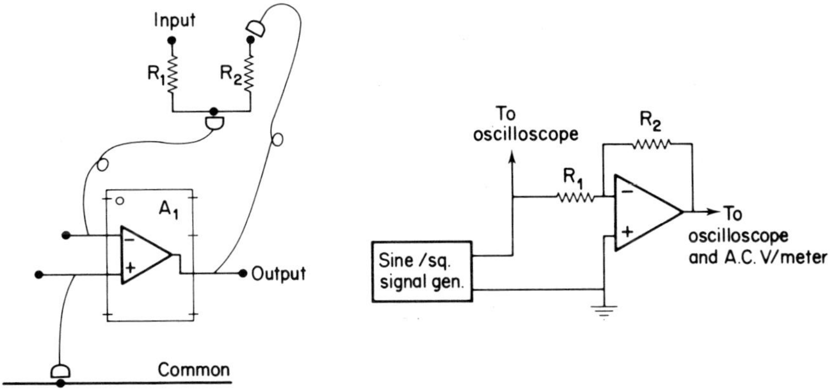

Figure 9 Inverting test circuit used for bandwidth and rise time measurements

Apply a sinusoidal input test signal, and monitor the input and output signals with a dual-trace oscilloscope. Bandwidth measurements are facilitated by the use of a high-impedance a.c. millivoltmeter (with dB scale). The voltmeter is first connected so as to read the input signal (in dB) which is applied to the test circuit. The signal generator is switched through its frequency ranges to ensure that this input signal amplitude remains constant as its frequency is varied. The voltmeter is then transferred to read the output of the test circuit (in dB).

Gain of test circuit (in dB) = output reading (in dB) − input reading (in dB)

A bandwidth measurement is performed quite simply by increasing the frequency of the test signal and noting the frequency at which the output of the test circuit falls by 3 dB (on its value at low frequencies).

As in the previous experiment the experimenter must ensure that the output signal amplitude is not so large as to cause slew rate limiting.

Rise time measurements are made by applying a square wave as the input signal to the circuit under test. The frequency of the test signal and the oscilloscope vertical and horizontal controls should be adjusted so as to obtain an output trace similar to that shown in Figure 10. The size of the output signal must not be so large as to cause any slew rate limiting.

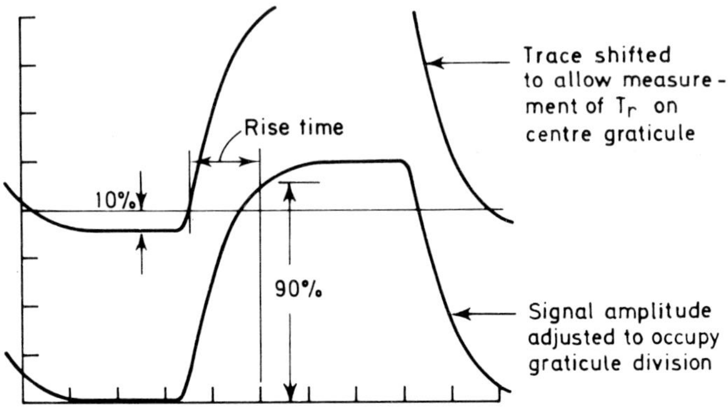

Figure 10 Display for rise time measurement (amplifier under test, type TL081; $R_1 = 1\ \text{k}\Omega$, $R_2 = 10\ \text{k}\Omega$; vertical sensitivity, 1 V/division; horizontal sensitivity, 1 μs/division)

The rise time of the circuit under test is the time taken for the output square wave to rise between 10% and 90% of its final value. A convenient procedure is to adjust the oscilloscope vertical sensitivity control so as to make the observed output square-wave trace occupy 5 graticule divisions. The rise time can be measured using the fine divisions on the central horizontal graticule line by suitably positioning the trace with the oscilloscope shift controls. The procedure is illustrated in Figure 10.

Values obtained for closed-loop bandwidth and rise time should be used as a check for the validity of equations 4.1 and 4.2. The value to be used for f_1, the unity-gain frequency of the amplifier, is that obtained in Experiment 3. Closed-loop frequency response plots should be superimposed upon the open-loop frequency plot obtained in Experiment 3 (Figure 8).

Further experimentation

In the above readings you may well observe what appear to be anomalous values at the higher settings of the feedback fraction β. In particular, if the experiment is performed with a TL081 type amplifier, the closed-loop frequency response graph, at values of β approaching units, may exhibit a rise before it rolls off (gain peaking). The transient response may exhibit overshoot and ringing. Both effects are indicative of an excess phase shift in the feedback loop causing an inadequate phase margin.

Excess phase shift can stem from two causes:

(a) Phase shift due to a second break in the amplifier open-loop frequency response roll-off; the second break occurs at a frequency which is a little higher than the unity-gain frequency f_1.
(b) Phase shift due to a frequency-dependent feedback fraction; stray capacity between the inverting input terminal of the operational amplifier and ground makes β frequency dependent (see Section 9.5 of companion text).

Measure the percentage overshoot in the transient response, and use the value that you obtain to estimate the damping factor (see Figure 3.19, Section 3.8.2, of companion text). Use this value of the damping factor to predict the gain peaking (see Figure 3.21 of companion text). Measure the gain peaking which actually occurs and compare with your estimated prediction.

Reduction in phase margin due to input capacity

Capacitance between the inverting input terminal of an operational amplifier and ground causes phase lag in the feedback fraction β and reduces phase margin. The phase variations in β alone can be made to control phase margin by choosing component values which make the $1/\beta$ and open-loop frequency response plots intersect at a frequency well removed from break frequencies in the open-loop response (see Figure 9.8, Section 9.5.2, of companion text).

The link connections shown in Figure 11 provide a test circuit which may be used to investigate the above effect. The amplifier is connected as an inverter, $R_2 = 100\ \text{k}\Omega$, $R_1 = 1\ \text{k}\Omega$, making

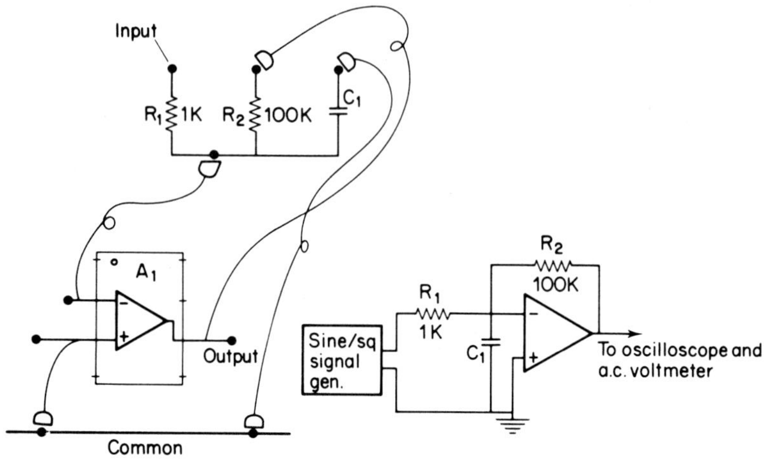

Figure 11 Test circuit used to investigate the effect of phase lag in the feedback fraction

$1/\beta = 101$. Capacitance is deliberately introduced at the inverting input terminal by connection of capacitor C_1.

Apply a square wave as an input test signal and observe the output signal of the circuit. Typical output transients observed with a TL081 type amplifier in the test circuit are shown in Figure 12. The lower

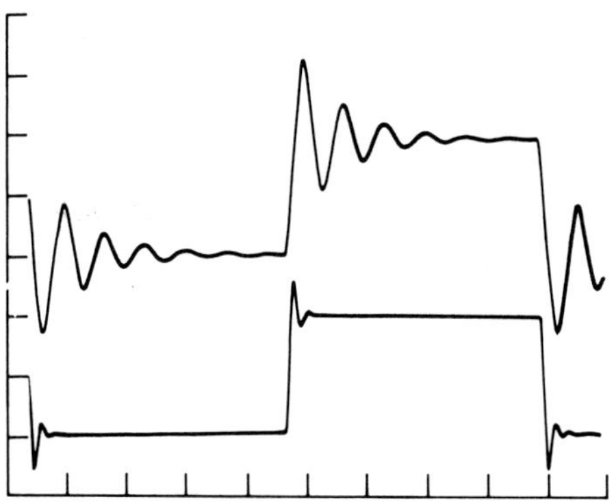

Figure 12 Output transients obtained with the test circuit of Figure 11 (amplifier under test, type TL081; top trace, $C_1 = 0.1\ \mu F$; bottom trace, $C_1 = 0.01\ \mu F$; vertical sensitivity, 1 V/division; horizontal sensitivity, 0.2 ms/division)

trace shows the output signal obtained with $C_1 = 0.01\ \mu F$; the upper trace shows the more lightly damped response obtained with $C_1 = 0.1\ \mu F$.

The percentage overshoot obtained in each case can be measured from the respective traces. Values of overshoot may be used to deduce the respective values of the damping factor (see Figure 3.19 of companion text). The amount of gain peaking to be expected in the sinusoidal response of the circuit can be predicted from damping factor values (see Figure 3.21 of companion text).

Measure the gain peaking by applying a sinusoidal input test signal and reading the output signal value on the dB scale of an a.c. voltmeter. The gain peak can be read directly from the dB scale by sweeping the frequency of the test signal. The gain peaking is simply the number of decibels by which the output goes up on its value at low frequencies.

It is instructive to measure the variation in closed-loop gain at frequencies in the neighbourhood of the peaking frequency. Results obtained with a TL081 type system are shown graphically in Figure 13. The graphs are shown in relationship to the open-loop roll-off and $1/\beta$ plot. Note that $1/\beta$ breaks up at the frequency

$$f_c = 1/2\pi C_1 R_1 \parallel R_2$$

C_1 was nominally $0.1\ \mu F$ but the results indicate a value of C_1 some 10% less than its nominal value.

The response marked A shows the variation in closed-loop signal gain of the inverting circuit. The response marked B shows the variation in closed-loop gain with the circuit used as a follower (input signal applied to the non-inverting input terminal of the amplifier and R_1 returned to common). Note that in the case of the follower connection the closed-loop gain shows an initial rise which is evidence of the frequency dependence of $1/\beta$. The closed-loop gain of a follower is given by

$$\frac{1}{\beta}\left(\frac{1}{1 + 1/\beta A_{OL(jf)}}\right)$$

Phase lag in the feedback fraction due to capacitance between the amplifier's inverting input terminal and power supply common can be compensated by connecting a capacitor in parallel with the feedback resistor R_2 (see Section 9.5.2, page 355, of companion text).

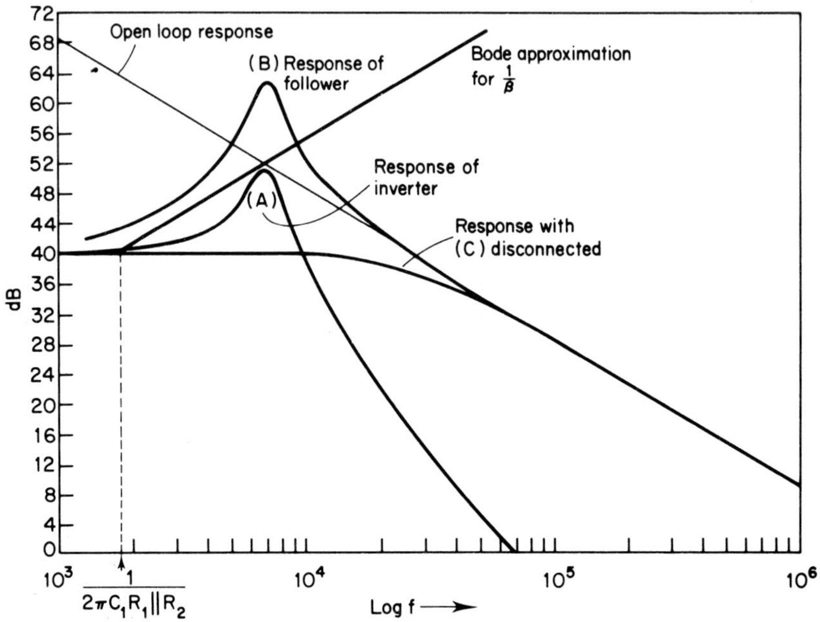

Figure 13 Closed-loop gain peaking due to phase lag in feedback fraction

Such a capacitor connection introduces a phase lead in the feedback fraction cancelling out the phase lag due to the input capacitance. Find a suitable lead capacitor value to compensate for the 0.1 μF, input capacitor; connect it in parallel with R_2 and observe the effect that this has on the dynamic response of the circuit.

Exercises 4

1 If the unity-gain frequency of an operational amplifier is 3×10^6 Hz, what will the small-signal closed-loop bandwidth and rise time be when the amplifier is connected as
(a) an inverting amplifier with signal gain = -1,
(b) a follower with gain 10,
(c) an inverting adder producing an output $V_o = -(V_1 + V_2)$

2 An operational amplifier has a capacitance (including stray wiring capacitance) of 10 pF between its non-inverting input terminal and power supply common. The amplifier is internally frequency compensated and has a unity-gain frequency 3×10^6 Hz; it is connected as a unity-gain inverter with input and feedback resistors both 100 kΩ. The circuit is found to exhibit gain peaking and overshoot in its transient response. Explain the reason for this lightly damped response and estimate the phase margin in the circuit. Suggest a method for increasing the phase margin in order to prevent the gain peaking.

Experiment 5

Slew rate, full power response frequency, power bandwidth

Parts of the companion text which are relevant to this experiment are Sections 2.8.2 and 2.9.

The slew rate of an operational amplifier is the maximum rate at which its output voltage can change. Because an amplifier has a slew rate limit it also has a maximum frequency at which it can produce a sinusoidal output signal with amplitude spanning the full output voltage range of the amplifier. If this frequency limit is exceeded, output waveform distortion occurs. The frequency limit is called the full power response frequency of the amplifier, f_p.

A sinusoidal signal of frequency f and peak-to-peak value V_{pp} has a maximum rate of change, $|dV/dt|_{max} = 2\pi f V_{pp}/2$. The maximum rate of change occurs as the sine wave passes through zero. The condition for no waveform distortion due to slew rate limiting is

$$2\pi f V_{pp}/2 \leqslant \text{slew rate}$$

A graph may be drawn relating the maximum peak-to-peak sinusoidal output voltage (achievable without slew rate distortion) to frequency. The graph shows what is sometimes called the power bandwidth of the amplifier. The power bandwidth of an amplifier may be less than the small-signal closed-loop bandwidth which is obtained when the amplifier is used in a negative feedback amplifier configuration. Closed-loop small-signal bandwidth, as we have seen from the previous experiment, depends upon the unity-gain frequency of the amplifier and the feedback fraction β which is set by the feedback components.

A value for amplifier slew rate may be obtained by measuring the

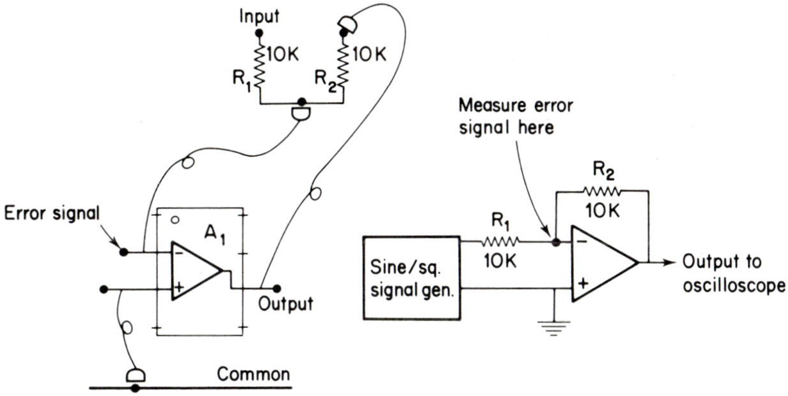

Figure 14 Test circuit for slew rate and full power response measurement

maximum rate of change of amplifier output voltage directly from an oscilloscope trace. A suitable test circuit is given in Figure 14; the amplifier under test is connected as a unity-gain inverter.

A square wave is used as an input test signal. Initially the amplitude of the test signal is set at a low value. The frequency of the test square wave and the oscilloscope controls are adjusted so as to display the closed-loop small-signal rise time of the circuit. The lower trace in Figure 15 shows the type of display required.

The amplitude of the test square wave is now increased. At first the displayed rise time remains unchanged, but when the rate of change of the output signal reaches the amplifier slew rate limit, the small-signal rise time of the circuit is no longer attained. The observed trace goes into a linear slew-rate-limited rate of change. The upper trace given in Figure 15 shows the type of behaviour to be expected. The slope of the linearly changing parts of this trace (expressed in V/μs) represents the slew rate of the amplifier.

The test circuit in Figure 14 can also be used for a measurement of the full power response frequency f_p of the amplifier. A sinusoidal test signal is used. The test signal amplitude is adjusted so that the amplifier produces a full-scale output sinusoid with no clipping of the peaks. The adjustment is made at a low frequency. The frequency of the test signal is then increased and the full power response frequency of the amplifier is measured as the frequency at which output waveform distortion first occurs.

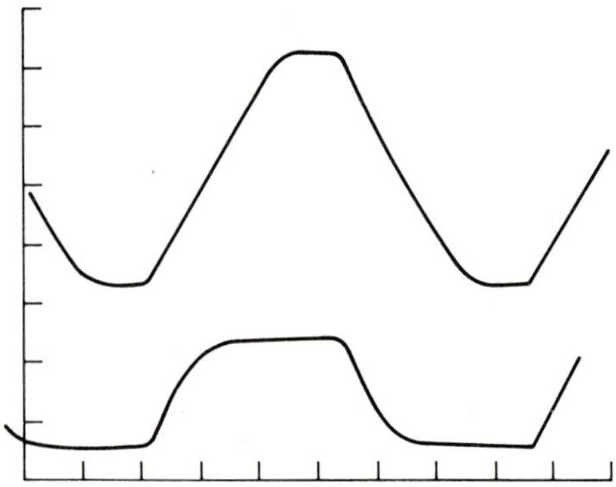

Figure 15 Oscilloscope traces obtained in slew rate measurement (amplifier under test, type 741); upper trace, slew-rate-limited rise (vertical sensitivity, 0.5 V/division; horizontal sensitivity, 2 μs/division); lower trace, small-signal rise time (vertical sensitivity, 0.2 V/division; horizontal sensitivity, 2 μs/division)

In practice it is not possible to detect a small amount of distortion in a sine wave by simply observing its waveform. A direct measurement of distortion requires the use of some form of waveform distortion meter. There is, however, an alternative method available if the amplifier under test is a type which has a full power response frequency which is appreciably less than the small-signal closed-loop bandwidth of the test circuit (e.g. a 741 type).

The alternative method of detecting the onset of output waveform distortion consists in observing the onset of appreciable distortion in the waveform of the signal which occurs at the inverting input terminal of the amplifier under test. This signal is an error signal, and it represents the difference between the externally applied test signal and the amplifier output signal. At frequencies appreciably less than the small-signal closed-loop bandwidth of the test circuit the error signal is small. Any differences between the output and input signal wave shape of the circuit causes a marked anomaly in the error signal waveform.

Examples of oscilloscope traces obtained in a test of this kind are shown in Figure 16. The amplifier under test was a 741 type. The upper trace in Figure 16 shows a full-scale output sine wave; no

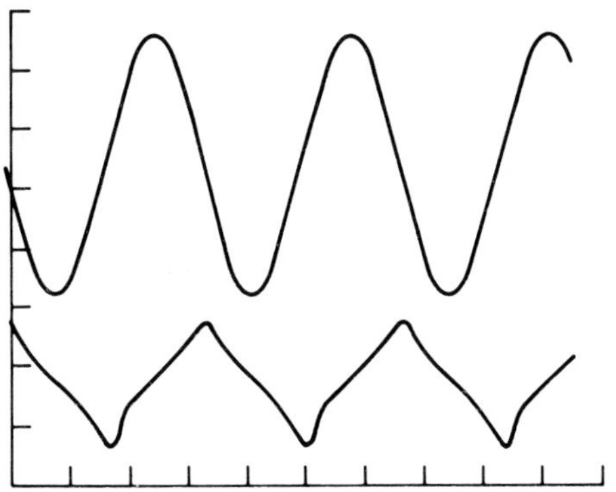

Figure 16 Distortion of the input error signal indicates onset of slew rate distortion in the output signal: upper trace, output signal (vertical sensitivity, 5 V/division; horizontal sensitivity, 50 μs/division); lower trace, input error signal (vertical sensitivity, 0.2 V/division; horizontal sensitivity, 50 μs/division)

distortion is evident in this trace. The lower trace shows the signal at the inverting input terminal of the amplifier under test. In this trace there is a quite marked distortion, which occurs as the output trace passes through zero. Slew rate distortion of the output wave, although undetectable in the output trace, causes the marked distortion in the input error signal.

In the case of sinusoidal output signals with amplitude less than the amplifier full-scale range, slew rate distortion occurs first at frequencies higher than f_p. A graph showing the power bandwidth of the amplifier can be drawn from values of the maximum frequency without distortion obtained for different values of output signal amplitude.

Exercises 5

1 An operational amplifier has an output slew rate limit of 0.5 V/μs. What is the highest frequency of a sinusoidal output signal which the amplifier can produce without slew rate distortion if the signal has a peak-to-peak value of (a) 20 V, (b) 10 V, (c) 5 V?

2 An operational amplifier has a unity-gain frequency of 3×10^6 Hz and a slew rate of 10 V/μs. What is the smallest value of the closed-loop gain ($1/\beta$) for which the power bandwidth of the circuit (20 V peak-to-peak output) will be no less than the small-signal bandwidth?

Experiment 6

Uncompensated amplifiers. Frequency compensation

Frequency-compensating techniques are discussed in Section 2.7 of the companion text.

Operational amplifiers which are classed as being externally frequency-compensated types need extra components connected to them externally to ensure the closed-loop stability of any negative feedback circuit in which they might be used. Adjusting the frequency–phase characteristics of a feedback loop in order to provide the loop with an adequate phase margin against instability is referred to as frequency compensation or phase compensation of the loop. In many cases frequency compensation of a circuit which uses an uncompensated amplifier is accomplished by simply connecting a single frequency-compensating capacitor between two of the external pins of the amplifiers. The effect that the size of this frequency-compensating capacitor has on circuit performance is investigated in this experiment.

The TL080 type amplifier is suggested as the test amplifier. It is an externally compensated version of the TL081 amplifier used in previous experiments. Several small-value capacitors, not included on the experimental board, are required in the experiment. Suggested nominal values are 5, 10, 20, 50, 100 and 500 pF. The capacitors should have leads which are long enough to allow them to be temporarily connected by wrapping their leads around circuit board pins.

Unity-gain frequency compensation

Connect the test amplifier as a unity-gain follower; the link connections are shown in Figure 17. Apply a square-wave input test signal with a peak-to-peak value of the order of 1 V and a frequency of say 200 kHz. Find the minimum value of the frequency-compensating capacitor (about 20 pF) which must be connected so that the output

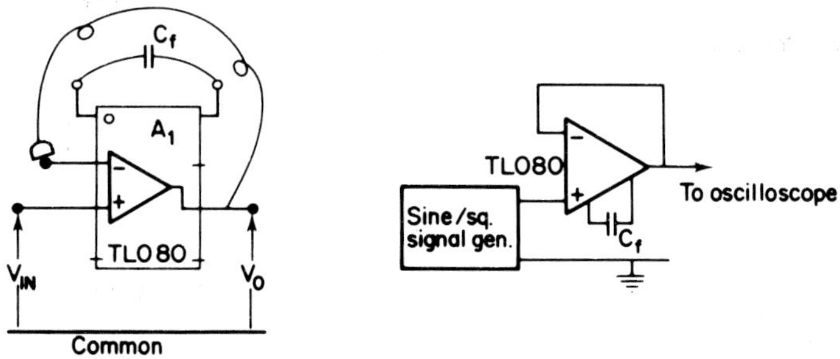

Figure 17 Unity-gain test circuit used to investigate the effect of frequency-compensating capacitor value

square wave exhibits no more than say 5–10% overshoot (output transients of the kind shown previously in Figure 12 are to be expected). Increase the amplitude of the test signal and measure the amplifier output slew rate (as in Experiment 5).

The TL080 amplifier with the frequency-compensating capacitor value determined by this experiment connected to it has performance characteristics almost identical to those of the internally frequency-compensated TL081 type.

Use of frequency-compensating capacitor values larger than that required for unity-gain compensation

Connect values of the frequency-compensating capacitance larger than that required for unity-gain frequency compensation. In each case measure the rise time, small-signal closed-loop bandwidth and slew rate of the unity-gain follower (as in Experiments 4 and 5).

Relate your results to the discussion given in Section 2.7 of the companion text.

Use of frequency-compensating capacitor values smaller than that required for unity-gain frequency compensation

An externally frequency-compensated amplifier, when used at closed-loop signal gains greater than unity, is closed-loop stable with a frequency-compensating capacitor value less than that required for unity-gain frequency compensation. In such cases the smaller frequency-compensating capacitor gives the amplifier a wider closed-loop bandwidth and a faster slewing rate than a similar internally frequency-compensated amplifier.

Connect the test amplifier as an inverter (see Figure 4). Use the values $R_2 = 10\ \text{k}\Omega$, $R_1 = 1\ \text{k}\Omega$, and connect 10 pF as a frequency-compensating capacitor. Measure the rise time, bandwidth and slew rate of the system (as in Experiments 4 and 5).

Change the value of R_2 to 100 kΩ and the frequency-compensating capacitor to 5 pF, and repeat the experiment.

Feed-forward frequency compensation

When used in the inverting feedback configuration at the higher values of closed-loop gain, a TL080 amplifier can be frequency compensated by a technique known as feed-forward frequency compensation. In this technique, the first internal gain stage in the amplifier is bypassed at the higher frequencies by a capacitor connected between the inverting input terminal of the amplifier and the input to the second internal gain stage (see Section 2.7.2 of companion text).

Make the connections shown in Figure 18. Measure the rise time, bandwidth and slew rate of the circuit (as in Experiments 4 and 5).

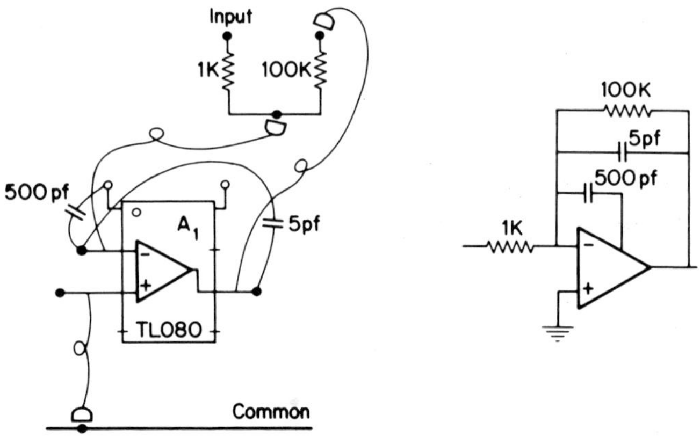

Figure 18 Feed-forward frequency compensation of a type TL080 amplifier

Exercises 6

1 A frequency-compensating capacitor of value 20 pF is required for unity-gain frequency compensation of an externally compensated FET input operational amplifier; with this value of frequency-compensating capacitor connected to it the amplifier has a unity-gain frequency 3×10^6 Hz and a slew rate 10 V/μs. It is required to use the amplifier at a closed-loop gain of 10 and to restrict the closed-loop bandwidth to a value 10 kHz by using a larger than normal frequency-compensating capacitor. What value frequency-compensating capacitor should be used? What will the amplifier slew rate be when this value frequency-compensating capacitor is connected to it?

2 A frequency-compensating capacitor of value 40 pF is required for unity-gain frequency compensation of an externally compensated bipolar input operational amplifier. The amplifier has a unity-gain frequency 800 kHz and a slew rate 0.4 V/μs when this capacitor is connected to it. The amplifier is used as an inverting adder with two input resistors each of value 10 kΩ and a feedback resistor value 10 kΩ. What is the minimum value of the frequency-compensating capacitor required by the amplifier when it is used

in this circuit? What will the small-signal bandwidth and slew rate of this circuit be when this minimum-value frequency-compensating capacitor is connected to the amplifier?

Experiment 7

Differential input amplifier configurations. Instrumentation amplifiers

Differential input amplifiers with a defined gain, a high input impedance and a high common mode rejection ratio (CMRR) are often used for the measurement of low-level signals. They are variously referred to as 'instrumentation amplifiers', 'measurement amplifiers' or 'data amplifiers'. The high CMRR of an instrumentation amplifier is put to advantage in rejecting the unwanted common mode interference signals which are often a major problem in low-level signal measurement.

Operational amplifiers are not classed as instrumentation amplifiers because they do not have a defined gain when used open loop. However, instrumentation amplifiers can be made by using operational amplifiers in appropriate negative feedback configurations. This experiment is concerned with examining the performance of such configurations. The circuits examined are discussed in Section 4.4 of the companion text.

A single-operational-amplifier differential input circuit

Figure 19 shows the link connections necessary to arrange a single operational amplifier as a defined-gain differential input amplifier. Signals applied to the inverting input of the circuit appear at the output of the amplifier multiplied by $-R_2/R_1$. Signals applied at the non-inverting input of the circuit appear at the output multiplied by R_2/R_1. The circuit has a differential gain R_2/R_1.

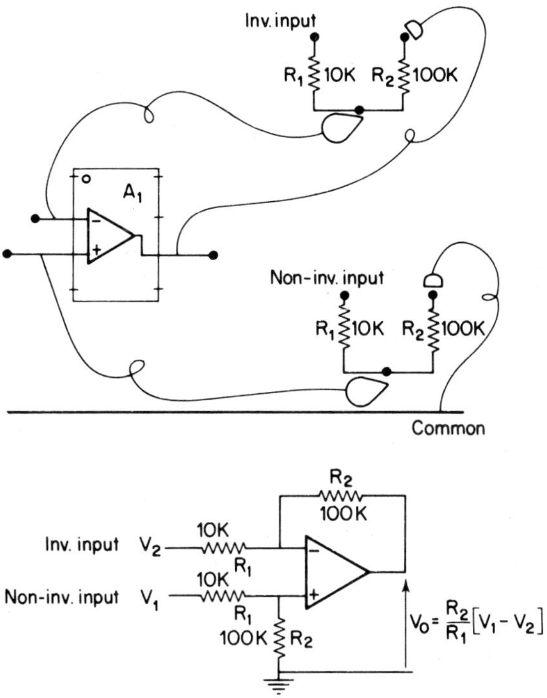

Figure 19 A differential input amplifier using a single operational amplifier

Apply a test signal to one input of the circuit and connect the other input to the common line; measure the gain and bandwidth as in Experiment 4.

Connect the two inputs together and apply the same *low-frequency* test signal to both inputs simultaneously as a common mode input signal. The ratio (output signal amplitude)/(common mode input signal amplitude) gives the common mode gain of the circuit. The ratio (differential gain)/(common mode gain) is the CMRR of the circuit.

The common mode gain should theoretically be zero and the CMRR infinite. The observed finite value for CMRR is largely due to mismatch in the resistor ratios.

Connect potentiometer P_2 in series with one of the R_2 resistors and find how big a value you can obtain for the CMRR by trimming this resistor. If you observe the phase of the common mode output signal relative to the phase of the common mode input signal for the untrimmed circuit, this will indicate which R_2 resistor you need to increase.

Observe the degradation in CMRR which occurs as the frequency of the common mode input signal is increased.

A high-input-impedance differential input circuit using two operational amplifiers

The link connections needed to realize a high-input-impedance differential input amplifier are shown in Figure 20. The circuit employs two operational amplifiers used in the follower configuration but cross-coupled together so as to obtain a differential input operation. A modification to the basic circuit which allows the gain of the circuit to be varied with a single potentiometer is also shown in Figure 20.

Measure the gain, bandwidth and CMRR of the basic circuit using an experimental procedure similar to that adopted for the one-amplifier differential input circuit of the previous section. As in the case of the single-amplifier circuit the CMRR is very much dependent upon resistor ratio matching. Try the effect of trimming the value of the appropriate R_2 resistor in order to obtain a high CMRR. Observe the degradation in CMRR which occurs as the frequency of the common mode input signal is increased.

Try the variable-gain modification. Check the validity of the variable-gain equation which is given for the circuit. This may be accomplished by fixing a measured gain and then measuring the value set for R_3 using a resistance meter.

Exercises 7

1 Sketch circuit arrangements for the following:

(a) a single-operational-amplifier circuit with the performance equation $V_o = V_1 - V_2$;

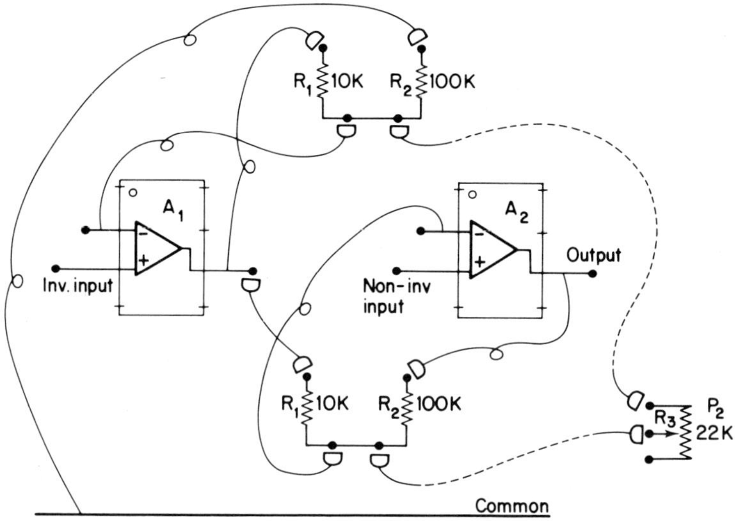

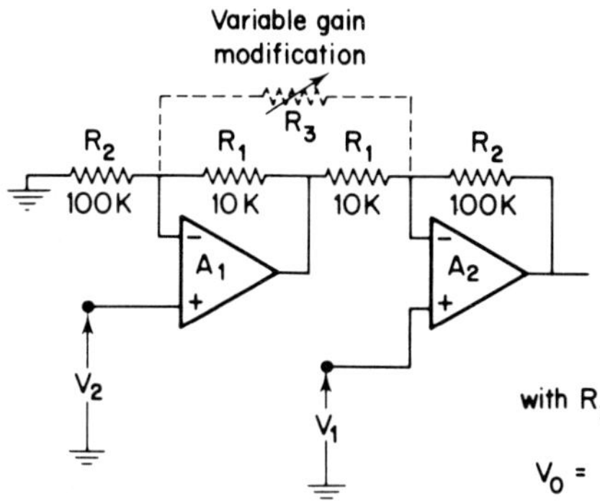

$$V_0 = \left[1 + \frac{R_2}{R_1}\right]\left[V_1 - V_2\right]$$

with R_3 in circuit

$$V_0 = \left[1 + \frac{R_2}{R_1} + \frac{2R_2}{R_3}\right]\left[V_1 - V_2\right]$$

Figure 20 A high-input-impedance differential input amplifier using two operational amplifiers

(b) a high-input-impedance two-operational-amplifier circuit with the performance equation $V_o = 2(V_1 - V_2)$.

Connect the circuits on the experimental board and verify their performance equations.

Experiment 8
Log and antilog circuits

An operational amplifier can be made to produce a defined non-linear input–output relationship. Non-linearity is achieved by connecting a defined non-linear circuit element in the feedback loop of the amplifier. Logarithmic (log) amplifiers generally employ silicon planar transistors as non-linear feedback elements.

Transistor logging equation

The short-circuit collector current (collector maintained at the same potential as the base) of a bipolar transistor is determined by the relationship

$$I_c = I_o\left[\exp\left(-\frac{q}{kt}V_{EB}\right) - 1\right] \quad (8.1)$$

where q is the electronic charge, k is Boltzmann's constant and V_{EB} is the emitter–base voltage. I_o is a constant at constant temperature (its value is usually of the order 10^{-13} A for silicon planar transistors). I_o approximately doubles for every 10 °C rise in temperature. T is the absolute temperature. The signs in equation 8.1 are those appropriate for an n.p.n. transistor. An n.p.n. transistor requires a negative emitter–base voltage for forward biassing of the emitter–base junction. Equation 8.1 reduces to

$$I_c = I_o \exp\left(-\frac{q}{kT}V_{EB}\right) \quad (8.2)$$

for I_c greater than 10^{-10} A. Equation 8.2 expresses a logarithmic relationship between collector current and emitter–base voltage. It may be written as

$$V_{EB} = -\frac{kT}{q}\log_e\left(\frac{I_c}{I_o}\right) \tag{8.3}$$

The main difficulty associated with the use of a single transistor as a logging element lies in the marked temperature dependence of the I_o term. However, if two transistors have their emitter terminals connected together the difference between their emitter–base voltages may be written as

$$V_{EB_2} - V_{EB_1} = \frac{kT}{q}\log_e\left(\frac{I_{c_1} I_{o_2}}{I_{c_2} I_{o_1}}\right) \tag{8.4}$$

If the transistors are both maintained at the same temperature, I_{o_1}/I_{o_2} remains constant and $V_{EB_2} - V_{EB_1}$ is made insensitive to the temperature dependence of I_{o_1} and I_{o_2}. It is this principle which underlies the log circuits which are examined in the present experiment. A full discussion of log and other non-linear circuits is given in Chapter 5 of the companion text.

Transistor log module

A matched pair of transistors and other components not included on the basic experimental board are required for the log circuit experiments. The interconnections between these extra components and their printed circuit board layout are shown in Figure 21. The printed circuit constitutes a log module which is link connected to the main experimental board.

Temperature-compensated log-of-voltage converter (see Section 5.5.1 of companion text)

A circuit for a temperature-compensated log-of-voltage converter is given in Figure 22. The link connections between the experimental board and the log module which are required to assemble the circuit are also given.

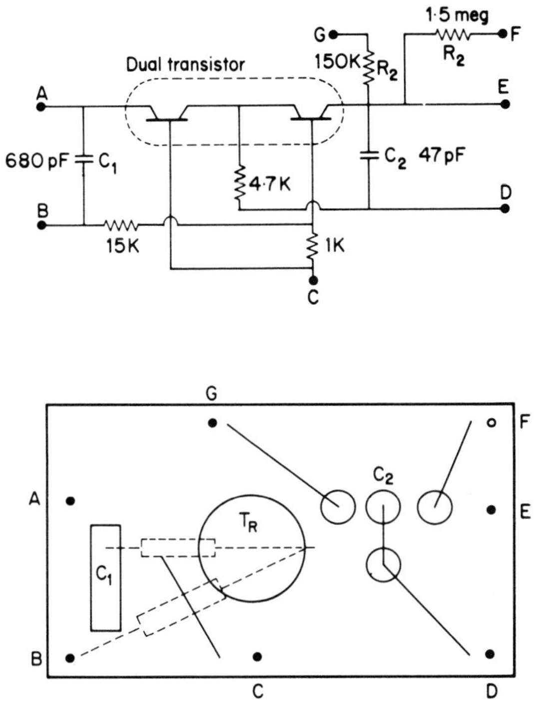

Figure 21 Circuit and physical layout of log module

Referring to the circuit diagram, the output voltage of amplifier A_1, attenuated by the resistive divider R_3, R_4, supplies a differential emitter-base voltage between transistors T_1 and T_2:

$$V_o \frac{R_3}{R_3 + R_4} = V_{EB_1} - V_{EB_2} \tag{8.5}$$

Use of equation 8.4 and rearrangement gives an expression for the output voltage of A_1 as

$$V_o = -\frac{R_3 + R_4}{R_3} \frac{kT}{q} \log_e \left(\frac{I_{c_1}}{I_{c_2}} \frac{I_{o_2}}{I_{o_1}} \right) \tag{8.6}$$

The circuit of Figure 22 sets $I_{c_1} = V_{in}/R_1$ and $I_{c_2} = V_s^+/R_2$. Substi-

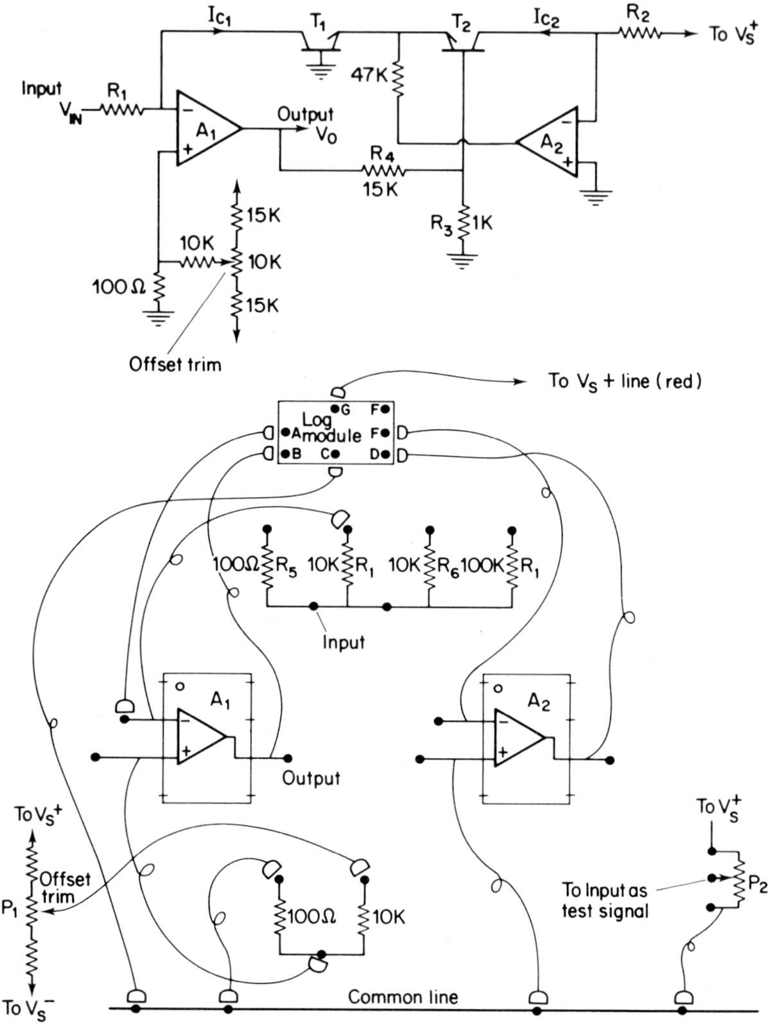

Figure 22 A temperature-compensated log-of-voltage converter

tution and expressing logarithms to the more usual base 10 gives the circuit performance equation as

$$V_o = -2.3\frac{kT}{q}\frac{R_3 + R_4}{R_3}\log_{10}\left(\frac{V_{in}}{V_s^+ + \dfrac{R_1 I_{o_1}}{R_2 I_{o_2}}}\right)$$

which may be written as

$$V_o = -K\log_{10}(V_{in}/V_{ref}) \tag{8.7}$$

The scaling factor is

$$K = 2.3\frac{kT}{q}\frac{R_3 + R_4}{R_3} \tag{8.8}$$

$kT/q \approx 60$ mV at 23 °C, and the values used for R_3 and R_4 make $K \approx 1$ V. The scaling factor exhibits a linear temperature dependence; this can be compensated by using a temperature-sensitive resistor for R_3.

$$V_{ref} = \frac{V_s^+ R_1}{R_2}\frac{I_{o_1}}{I_{o_2}} \tag{8.9}$$

Measurements should be made to find the range of input signal values (the logging range) for which equation 8.7 is a valid representation of circuit behaviour.

Experimental procedure

No offset trim

Make the link connections shown in Figure 22. Initially it is suggested that the offset trimming arrangement be omitted. The non-inverting input terminal of amplifier A_1 should be linked directly to the power supply common line. Select a value $R_1 = 10$ kΩ on the main experimental board and a value $R_2 = 150$ kΩ on the log module.

Starting with V_{in} of say 15 V (and reducing), apply measured input signals and find the corresponding values of the output signal produced. Input signals are conveniently derived from potentiometer P_2 connected between the positive supply line and the power supply common line. The lower values of input signals can be obtained by attenuating the signal from P_2 using resistors R_5 and R_6 as a potential divider (connections are indicated in Figure 22). Results should be plotted on a piece of log–linear graph paper as V_o versus log V_{in}. Continue to reduce V_{in} until a marked deviation from the logging relationship is observed.

The lower limit to the logging range of the converter is determined by the offset errors of amplifier A_1. Offset errors in amplifier A_2 do not significantly affect the logging range; they have a small, but insignificant, effect on V_{ref}.

Offset trim extends the lower limit of the logging range

When an operational amplifier is connected in a log converter circuit, considerations regarding the trimming of its offset error are somewhat different from when the amplifier is used in a linear feedback configuration. It is necessary to examine the implications of the performance equation of the circuit (equation 8.7).

A log circuit *does not give zero output for a zero input signal.* Output depends upon the logarithm of the ratio V_{in}/V_{ref}. Output is zero when $V_{in} = V_{ref}$ ($\log_{10} 1 = 0$). When $V_{in} = \frac{1}{10}V_{ref}$ the output is $V_o = -K\log(1/10) = +K$. When $V_{in} = \frac{1}{100}V_{ref}$, $V_o = -K\log(1/100) = +2K$. As the input signal is reduced, the output goes increasingly more positive. A zero input signal theoretically gives rise to an infinite output signal.

In practice, as the externally applied input signal approaches zero the output signal takes on a value which is determined by the input offset error of amplifier A_1, E_{os}:

$$E_{os} = \pm V_{io} - I_b R_1$$

$$\text{When } V_{in} = 0,\ V_o = -K\log_{10}(E_{os}/V_{ref}) \tag{8.10}$$

Equation 8.10 serves as a basis for an offset trimming procedure.

A TL081 amplifier has an input offset voltage temperature coefficient specified as typically 10 μV/°C. It would therefore seem feasible to trim out the offset error of amplifier A_1 to within say 30μV. The bias current contribution to E_{os} is negligible with the particular value used for R_1.

Offset trim procedure

Add the link connections (shown in Figure 22) required to apply an offset to the non-inverting input terminal of amplifier A_1. Set $V_{in} = 0$ and adjust P_1 to make $V_o = -K\log_{10}(3 \times 10^{-5}/V_{ref})$. With $V_{ref} =$ 1 V and $K \approx 1$ V the output should be adjusted to approximately 4.5 V.

After completion of the offset trim the logging characteristic of the circuit should be measured again. Further sets of measurements may be taken to demonstrate the dependence of V_{ref} on the values of resistors R_1 and R_2. A value $R_2 = 1.5$ MΩ is selected by connecting the power supply positive line to the point G on the log module (instead of point F). The resistor R_1 may be made 100 KΩ by appropriate link selection on the main experimental board.

Note that the output signal crosses zero when

$$V_{in} = V_{ref} = V_s^+ \frac{R_1}{R_2} \frac{I_{o_1}}{I_{o_2}}$$

Zero-crossing values may be used to deduce the extent of the

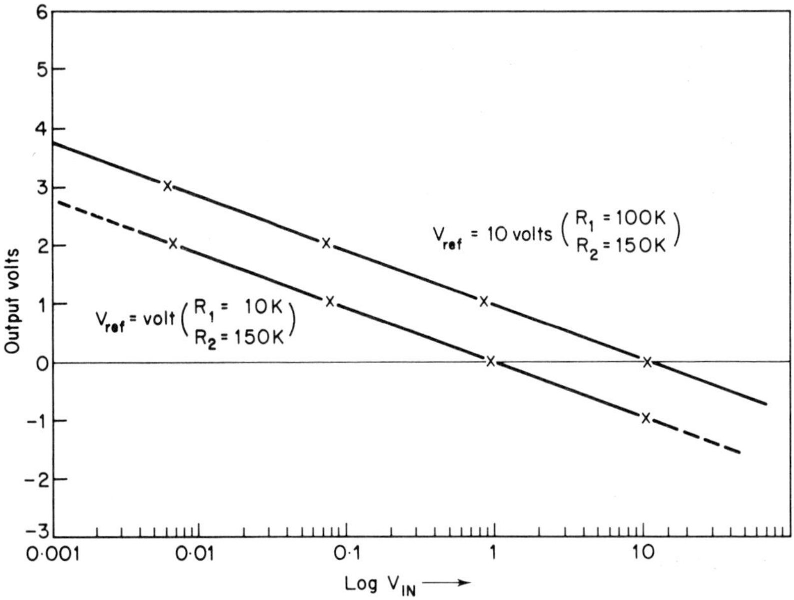

Figure 23 Logging characteristics obtained with a temperature-compensated log converter

matching in the transistor I_{o_1} and I_{o_2} terms. Logging characteristics are conveniently presented in the graphical form shown in Figure 23.

Consider the circuit parameters which influence the slope of the logging characteristics (the value of the scaling factor K). Measured slopes may be used to deduce which circuit component (R_3 or R_4) you could increase, using an externally connected trim potentiometer, in order to set the scaling factor to exactly unity (1 V/decade).

It is important to note that log converters can accept only single-polarity input signals. The circuit of Figure 22 can be used only for positive input signals, i.e. signals which cause an input current to flow towards the inverting input terminal of amplifier A_1. Negative input signals cause the output of amplifier A_1 to go into its positive saturation state.

Temperature-compensated antilog converter

The link connections between the experimental board and the log module can be modified in order to make a circuit which performs an antilog conversion. The circuit and the required link connections are given in Figure 24. The performance equation for the circuit is derived in Section 5.5.5. of the companion text. It is

$$V_o = M 10^{-V_{in}/K} \tag{8.11}$$

The multiplying factor M is related to circuit parameters by the equation

$$M = V_s^+ \frac{R_2 I_{o_2}}{R_1 I_{o_1}} \tag{8.12}$$

The factor K is determined as before as

$$K = 2.3 \frac{kT}{q} \frac{R_3 + R_4}{R_3} \tag{8.13}$$

Experimental procedure

Make the link connections shown in Figure 24. The values $R_1 =$ 150 kΩ, $R_2 = 10$ kΩ, $V_s^+ = 15$ V make the nominal value of the

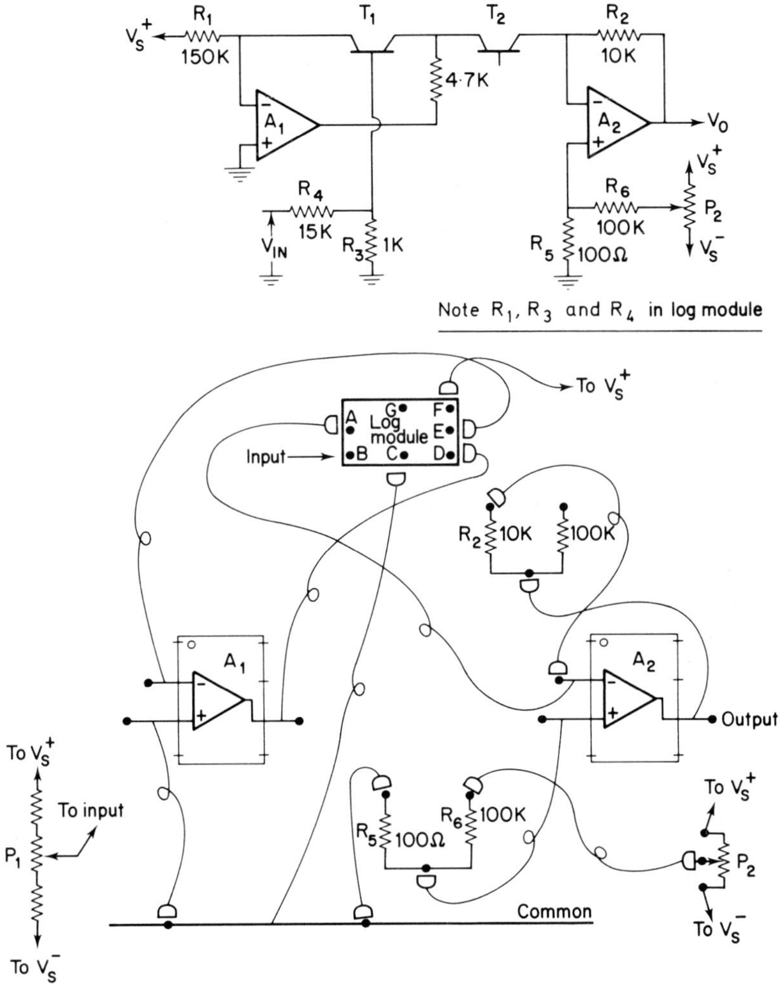

Figure 24 A temperature-compensated antilog converter

multiplying factor M unity. If required, resistors can be trimmed to make M exactly unity. This is done by setting $V_{in} = 0$ and adjusting resistor values to make the output voltage exactly equal to the value required for M. (Note $V_o = M10^0 = M$.)

Adjust the offset error of A_2. Set V_{in} sufficiently positive to cut off transistor T_1 (say 15 V), and adjust the offset trim potentiometer P_2 to make $V_o = 0$.

Now apply measured input signals and find the corresponding output voltage values produced. Hence determine the range of input signal values for which the performance equation (Equation 8.11) is valid. Both positive and negative input signals are allowed. Examination of the performance equation shows that negative input signals give rise to output voltages greater than M, while positive input signals produce output voltages less than M, when $V_{in} = 0$, $V_o = M$.

Resistors R_3 and R_4 are proportioned so as to make the factor K nominally unity. K can be set to exactly unity by suitably trimming component values. If required, this may be done by setting $V_{in} = -1$ V and adjusting the potential divider formed by R_3 and R_4 so as to make the output voltage $V_o = 10M$ V.

The antilog characteristics of the circuit are conveniently presented in the graphical form shown in Figure 25.

If two experimental boards and two log modules are available, log and antilog converters can be combined to make a multi-function unit capable of performing computations embodied in a performance equation of the form

$$V_o = V_{in_3}(V_{in_1}/V_{in_2})^n$$

where n is an arbitrary power set by the choice of component values. Readers interested in pursuing further experimentation with a unit of this kind are advised to consult Section 5.6.2. of the companion text.

Exercises 8

1 A log converter has the performance equation $V_o = -K\log_{10}(V_{in}/V_{ref})$.

(a) If $V_{ref} = 1$ V and $K = 2$, what output voltage does the con-

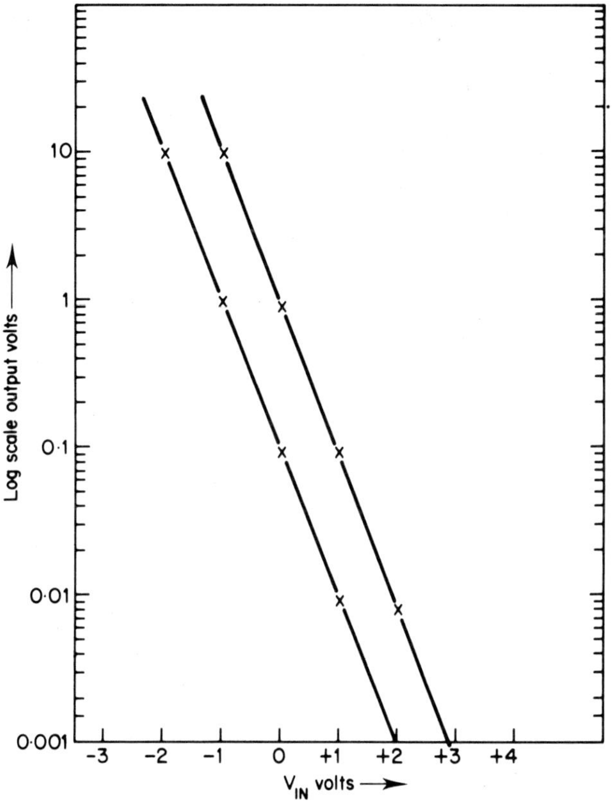

Figure 25 Antilog characteristics obtained with the circuit of Figure 24

verter produce when the input signal voltage is 10 mV, 100 mV, 1 V, 5 V, 10 V?

(b) What values of the scaling factor K and the reference voltage V_{ref} are required if input signals in the range 1 mV to 10 V are required to produce output signals in the range +8 V to −8 V?

2 The log converter of Figure 22 uses $R_1 = 10\,\mathrm{k\Omega}$, $R_2 = 150\,\mathrm{k\Omega}$, and $V_s^+ = 15$ V; zero output is produced by an input signal of 1.2 V. Deduce the extent of the mismatch in the logging transistors T_1 and T_2 (find the ratio I_{o_1}/I_{o_2}).

3 The input resistor R_1 in Figure 22 is removed and a high-impedance current source is connected so that it supplies an input current directly at the inverting input terminal of amplifier A_1. The circuit then acts as a log-of-current converter with the performance equation $V_o = -K\log_{10}(I_{in}/I_{ref})$. In what way are the scaling factor (K) and reference current (I_{ref}) values related to circuit parameters? Input currents in the range 10 nA to 1 mA are required to produce output voltages in the range +10 V to −10 V; find suitable values for K and I_{ref}. Assuming that $R_3 = 1$ kΩ, $V_s^+ =$ 15 V and $I_{o_1} = I_{o_2}$ find the values of R_4 and R_2 required to establish these values of K and I_{ref}.

Experiment 9
Integrator circuits

An operational amplifier with a capacitor connected between its output terminal and its inverting input terminal produces an output voltage which is proportional to the time integral of the current arriving at its inverting input terminal. This section shows how to arrange integrator circuits on the experimental board and suggests experimental procedures suitable for evaluating the performance of these circuits. The detailed theoretical and practical aspects which govern the performance of operational integrators are discussed in Sections 6.1–6.6 of the companion text.

A basic voltage integrator. Integrator drift

A circuit diagram for a basic voltage integrator is given in Figure 26; also shown are the link connections which are required to assemble the circuit on the experimental board. The ideal performance equation for the circuit is

$$V_o = -\frac{1}{CR}\int V_{in}\,\mathrm{d}t \qquad (9.1)$$

Experimental procedure

Make the link connections shown in Figure 26. Various integrator CR values can be selected but it is suggested that you use $C = 0.1\ \mu\text{F}$, $R = 1\ \text{M}\Omega$ initially. Also initially omit the offset trim arrangement

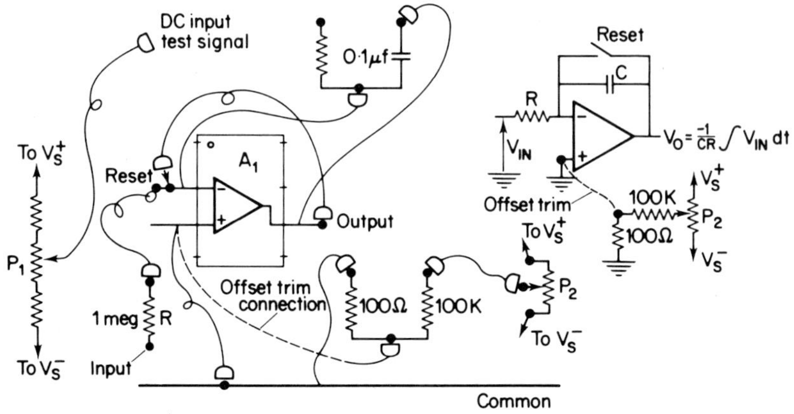

Figure 26 A basic voltage integrator circuit

which is shown. Set V_{in} at zero by connecting the input point of the circuit to the power supply common line. Set the output to zero by closing the reset switch. A flying link can be made to serve as a reset switch. Open the reset switch and measure the rate at which the output voltage changes. The measurement can be made using a voltmeter and stop-watch or an oscilloscope set to a slow sweep rate.

A practical voltage integrator does not give zero output when its input voltage is set at zero. Its output drifts with time. Integrator drift is due to the combined effects of the operational amplifier input offset voltage and bias current. The drift rate dV_o/dt is governed by the relationship

$$\frac{dV_o}{dt} = \pm\frac{V_{io}}{CR} + \frac{I_b}{C} \qquad (9.2)$$

If an FET input operational amplifier (a TL081) is used in the circuit, the bias current contribution I_b/C to the observed drift rate C is likely to be much less than the input offset voltage contribution V_{io}/CR. Assuming you have previously measured the value of the amplifier input offset voltage (Experiment 2), use this value as a check for the validity of Equation 9.2.

Offset trim

Connect the non-inverting input terminal of the operational amplifier to the offset trimming arrangement shown in Figure 26. Set V_{in} at zero and reset the output to zero as before. When the reset switch is opened the output drift rate of the integrator can be set at very low value by careful adjustment of potentiometer P_2. The adjustment in effect applies a voltage to the non-inverting input terminal of the operational amplifier which is equal but opposite to the input offset error of the amplifier ($E_{os} = \pm V_{io} + I_b R$). (Note that a zero output drift rate, established as a result of offset trimming, is unfortunately not maintained over the long term. Operational amplifier input offset voltage and bias current are both temperature dependent.)

Having performed the offset trim adjustment, apply measured input signals and observe the corresponding values of the output signal. A d.c. input test signal is conveniently derived from potentiometer P_1. If V_{in} is constant the integrator produces an output which changes linearly with time:

$$V_o = -\frac{1}{CR}\int V_{in}\,dt = -\frac{V_{in}}{CR}t \quad (V_{in}\ \text{constant})$$

A positive input gives a linear negative-going output ramp, while a negative input gives a positive ramp.

A.c integrator

In integrator applications in which input signals are alternating about a mean level of zero the use of a so-called a.c. integrator circuit can obviate the necessity for providing an integrator reset facility. A circuit for a basic a.c. integrator is shown in Figure 27. The difference between this circuit and the previous integrator circuit is simply the inclusion of the resistor R_2 connected in parallel with the feedback capacitor of the integrator.

Connect up the circuit given in Figure 27. Apply a sinusoidal input test signal and take measurements to find the way in which the gain varies with frequency. Also measure the way in which the phase relationship between input and output signals varies with frequency.

Note that at frequencies a decade higher than the frequency $f = 1/2\pi CR_2$ the gain–frequency plot should closely approximate an ideal integrator 20 dB/decade roll-off. The phase shift should closely approximate the ideal integration 90° phase shift. As the frequency is reducing towards the value $f = 1/2\pi CR_2$ the response may be expected to be increasingly in error from ideal integrator behaviour. At frequencies more than a decade below $f = 1/2\pi CR_2$ the circuit behaves as an inverter with gain $-R_2/R$.

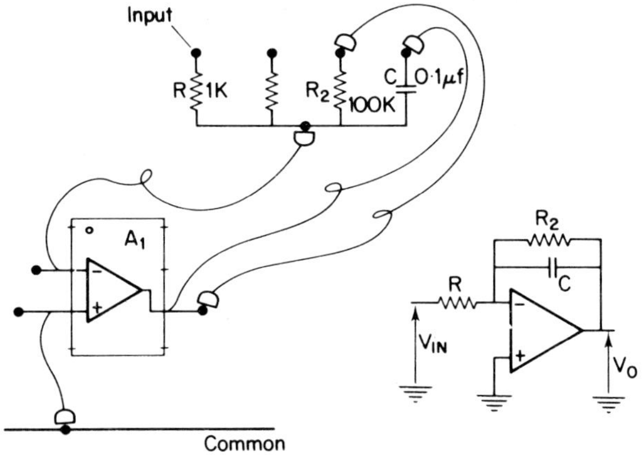

Figure 27 A basic a.c. integrator circuit

Apply a square wave as an alternative input test signal. Monitor input and output signals with a dual-trace oscilloscope. Try the effect of varying the frequency of the test signal and of selecting different integrator CR values. If the frequency f of the input square wave is such that $f \gg 1/2\pi CR_2$, the a.c. integrator will give a triangular output wave.

Current integrator

An FET input operational amplifier, arranged as a current integrator, can be used to measure very small currents (including its own bias current). A current integrator circuit is given in Figure 28.

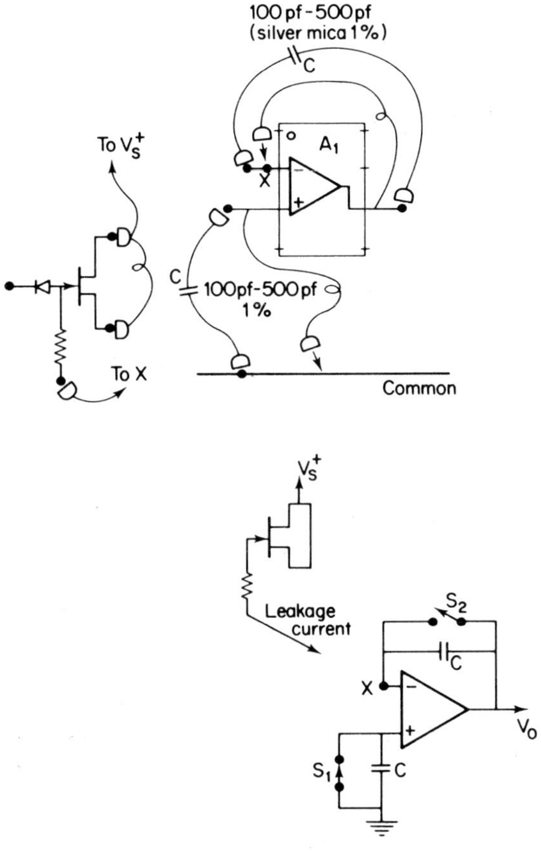

Figure 28 An FET input operational amplifier connected as a current integrator measures small currents

Experimental procedure

Make the link connections given in Figure 28. Use an oscilloscope to measure the output voltage produced by the amplifier; the time base of the oscilloscope should be set to a slow sweep rate. Start with switches S_1 and S_2 closed (link connections can serve as switches). Open S_2 and measure the rate of change of the amplifier output voltage (the slope of the oscilloscope trace). The observed output ramp is produced as a result of the continuous charging of the feedback capacitor by the bias current flowing to the inverting input terminal of the operational amplifier. The ramp rate (ramp up) is

$$\mathrm{d}V_o/\mathrm{d}t = I_b^-/C \tag{9.3}$$

If a close-tolerance capacitor is employed, the measured slope of the output ramp can be used to deduce a value for the bias current to the inverting input terminal of the operational amplifier.

Now close S_2 and open S_1. Bias current drawn by the non-inverting input terminal of the amplifier causes a continuous charging of the capacitor which is connected between the non-inverting input terminal of the amplifier and the power supply common. The output voltage of the amplifier ramps down. Measure the ramp rate. Use your measured value to deduce the bias current I_b^+ from the relationship $\mathrm{d}V_o/\mathrm{d}t = -I_b^+/C$.

If both S_1 and S_2 are opened the output drift rate becomes

$$\mathrm{d}V_o/\mathrm{d}t = (I_b^- - I_b^+)/C = I_{io}/C$$

The current integrator can be used to measure the small reverse leakage currents of semiconductor devices. Connections required to measure the gate leakage current of the FET on the experimental board are indicated in Figure 28. The drain and source of the FET are commoned and connected to the positive supply rail. The gate is connected to the operational amplifier summing point (point X). S_1 and S_2 are initially both closed. S_2 is then opened and the slope of the output ramp is measured:

$$\mathrm{d}V_o/\mathrm{d}t = (I_b^- - I_L)/C \tag{9.4}$$

with I_b^- having been previously measured, the leakage current I_L of the FET can be deduced. Note that if $I_L < I_b^-$ the output ramps up, while if $I_L > I_b^-$ the output ramps down.

The technique described above can also be used to measure the leakage currents of the diodes which are included on the experimental board.

Exercises 9

1 An operational amplifier has input offset voltage $V_{io} = 3$ mV and bias current $I_b = 50$ pA. The temperature coefficient of the input

offset voltage is $\Delta V_{io}/\Delta T = 10\ \mu V/°C$ and the bias current doubles for every 10 °C rise in temperature. The amplifier is connected as an integrator with $R = 10\ M\Omega$, $C = 0.1\ \mu F$ (Figure 26).
(a) What is the worst-case output drift rate with no offset trim?
(b) What is the worst-case output drift rate if the offset is initially trimmed for zero drift and the temperature then increases by 10 °C?

2 A ± 2 V peak-to-peak square wave of frequency 1 kHz is applied as an input signal to the a.c. integrator circuit of Figure 27 ($C = 0.1\ \mu F$, $R = 1\ k\Omega$, $R_2 = 100\ k\Omega$). What is the amplitude of the triangular-wave output signal? Estimate the percentage deviation from linearity of the triangular wave (use equation 6.10 of the companion text and set $A_o R' = R_2$).

3 An FET input operational amplifier has its non-inverting input terminal connected to power supply common and a capacitor value 100 pF is connected between its output terminal and its inverting input terminal. The capacitor is initially shorted; when the short is removed the output voltage of the amplifier ramps up at a rate 600 mV/s. Deduce the value of the bias current drawn at the inverting input terminal of the amplifier.

Experiment 10
A differentiator. Stability considerations

In this experiment the reader is asked to make the link connections which arrange an operational amplifier to act as a differentiator. The circuit is shown in Figure 29. The ideal performance equation for the differentiator circuit is

$$V_o = -CR\,\mathrm{d}V_{in}/\mathrm{d}t \qquad (10.1)$$

In a basic differentiator circuit, phase lag introduced into the feedback loop by the presence of the input capacitor C causes the circuit to have an inadequate phase margin. The practical consequences of this are examined in the experiment and a method of increasing the

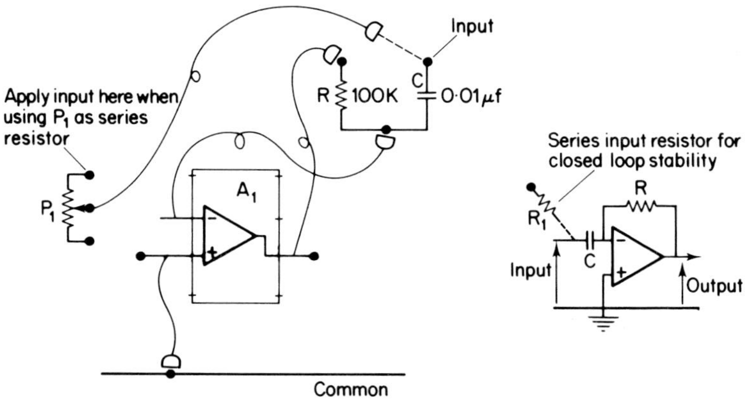

Figure 29 A differentiator circuit

phase margin is investigated. Differentiator circuits are discussed in Sections 6.7 and 6.8 of the companion text.

Experimental procedure

Make the link connections shown in Figure 29. An input test signal having a triangular waveform is a convenient one with which to examine circuit behaviour. The test signal should be applied to the input point of the circuit, and the input and output signals should be monitored with a dual-trace oscilloscope.

If a test signal generator providing a triangular waveform is not available a triangular wave can be formed by integrating a square wave. Amplifier A_2, on the board, can be connected as an a.c. integrator for this purpose. A_2 should be connected, as shown previously in Figure 27, with component values $R = 10\ k\Omega$, $C = 0.1\ \mu F$, $R_2 = 100\ k\Omega$, selected from the lower bank of components on the experimental board. A square wave applied to the A_2 integrator circuit gives a triangular output wave which can be used as an input test signal to the differentiator.

The triangular test signal should initially be applied directly to the input capacitor of the differentiator. The differentiator component values suggested in Figure 29 ($C = 0.01\ \mu F$, $R = 100\ k\Omega$) make a test signal with a frequency of approximately 500 Hz convenient for oscilloscope presentation. The upper and middle traces given in Figure 30 show typical input and output waveforms obtained in the test. The lightly damped output response is due to the phase lag in the feedback signal caused by the input capacitor C.

Now insert potentiometer P_1 as a resistor in series with the input capacitor (link connections are indicated in Figure 29). Adjust the value of P_1 in order to just prevent the overshoot and ringing in the differentiator output signal. The lower trace given in Figure 30 indicates the adjustment which is to be aimed for. Use a resistance meter to measure the value of the series resistance R_1 that has been introduced into the input path of the differentiator.

Measure the amplitudes and time periods of the upper and lower traces in Figure 30; the values should be used as a check on the validity of the differentiator performance equation (Equation 10.1). The lightly damped response indicated by the middle trace and the

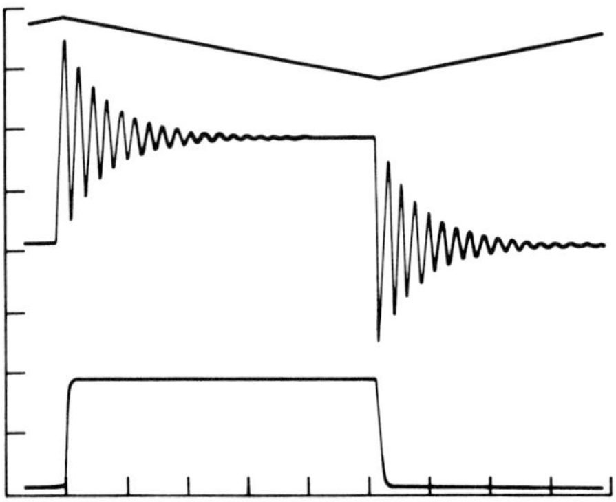

Figure 30 Waveforms obtained in a differentiator test: top trace, input; middle trace, output without series input resistor; bottom trace, output with series input resistor

value required for R_1 to prevent it should be interpreted in terms of the phase margin in the feedback loop of the circuit.

Phase margin may be assessed by sketching Bode plots for $1/\beta_{(jf)}$ and finding the frequency f_1^1 at which they intersect the open-loop frequency response plot of the operational amplifier. The form which these plots have for the basic differentiator and for the differentiator with series input resistor R_1 are shown in Figures 31a and 31b respectively.

The basic differentiator has its value for $1/\beta_{(jf)}$ determined by $1/\beta_{(jf)} = 1 + j2\pi fCR$. The phase shift ϕ in the loop gain at frequency f_1^1 is $\tan^{-1}(f_1^1/f_c)$, due to phase lag in the open-loop gain, plus $\tan^{-1}(f_1^1/f_c)$, due to phase lag in β.

The differentiator with series input resistor R_1 has

$$\frac{1}{\beta_{(jf)}} = \frac{1 + j2\pi fC(R + R_1)}{1 + j2\pi fCR_1} \qquad (10.2)$$

In Figure 31b the phase shift ϕ in the loop gain at frequency f_1^1 is $\tan^{-1}(f_1^1/f_c)$ in the open-loop gain, plus $\tan^{-1}(f_1^1/f_{c_1}) - \tan^{-1}(f_1^1/f_{c_2})$ in β. The phase margin is $\theta_m = 180 - \phi$.

The output resistance of a signal source used to supply the input signal of a differentiator constitutes a resistance in series with the input capacitor of the differentiator. The phase margin for a practical

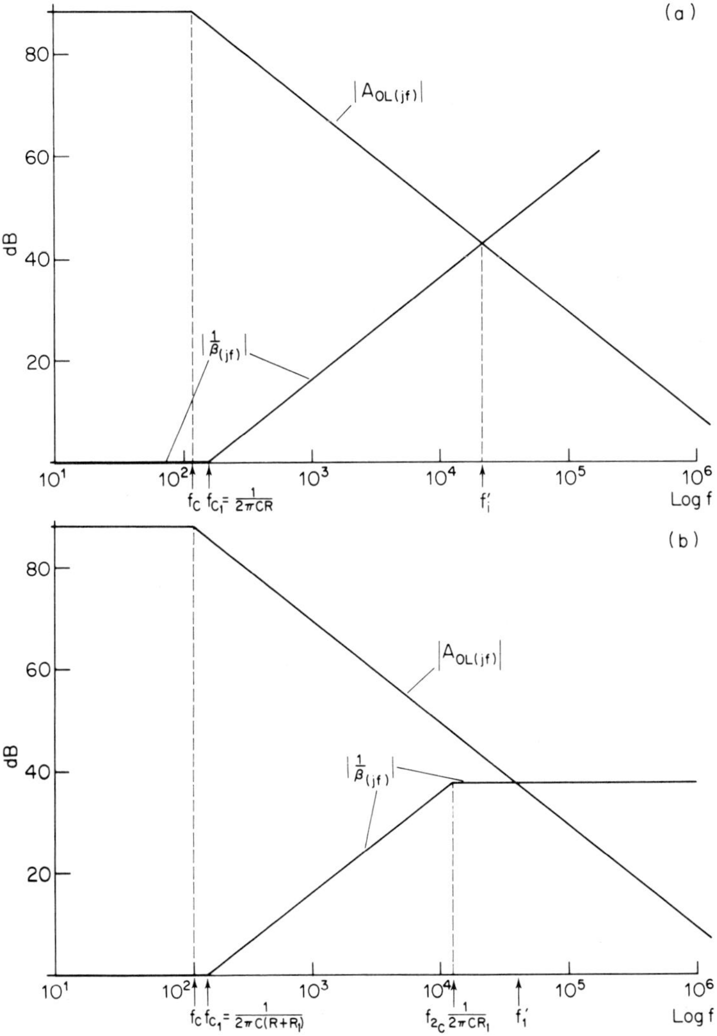

Figure 31 Differentiator Bode plots

differentiator circuit used without an added series resistor is therefore likely to be greater than that to be expected from the Bode plots in Figure 31a.

Exercises 10

1 A ± 5 V peak-to-peak triangular wave with frequency 250 Hz is applied as an input test signal to the differentiator circuit in Figure 29. What is the amplitude of the square wave which should be produced at the output of the differentiator?

2 An internally frequency-compensated operational amplifier with unity-gain frequency 3×10^6 Hz is connected as a simple differentiator circuit with component values $C = 0.01\ \mu F$, $R = 100\ k\Omega$. If the circuit is supplied by a signal source with output resistance 60 Ω, what do you expect to be the value of the phase margin in the circuit?
Estimate the value of series input resistor required in order to make the $1/\beta_{(jf)}$ plot break back two octaves before its intersection with the $|A_{OL(jf)}|$ plot.
With this value of series resistor connected in the circuit estimate the phase margin and the highest frequency for which the amplitude of the differentiator output signal will be in error by no more than 5% from its ideal value.

Experiment 11
Active filters

An electrical filter is a functional device which is used to separate sinusoidal signals of different frequency. A filter passes sinusoidal signals with frequencies which lie in a specified range but reduces (attenuates) or blocks off signals with frequencies which lie outside that range. Filters which use amplifiers together with resistor–capacitor networks are called active filters.

This section gives several active filter circuits which can be readily connected up on the experimental board. Experimental procedures are suggested which are suitable for investigating the performance of the circuits. A brief summary of filter terminology and the general equations which describe filter response are given. Observations made in the experiments should be related to these general response equations.

Filter terminology

'High-pass', 'low-pass', 'band-pass' and 'band-reject' are terms used to describe the attenuation characteristics of a filter. A high-pass filter passes signals with frequency above a certain limit frequency but attenuates signals with frequency below that limit. A low-pass filter passes frequencies below a limit but attenuates frequencies above that limit. Band-pass filters pass frequencies which lie between two limit frequencies but attenuate all other frequencies. Band-reject or notch filters attenuate frequencies in a range between two limits but pass all other frequencies outside that range.

Filters are classified as to their 'order'. A filter composed of linear circuit elements has its input and output signals related by a linear differential equation. The term 'filter order' derives from the order of this differential equation. The highest-order differential coefficient occurring in the differential performance equation of a filter specifies the order of the filter.

The performance of a filter is described mathematically by its transfer function. The transfer function of a filter is derived by using the well-established mathematical techniques of network theory to transform the differential equation relating the output and input signals of the filter in the time domain to an equation which relates them in the complex frequency variable s.

First-order filter functions

	First-order low-pass	*First-order high-pass*
Transfer function	$F_{(s)} = \dfrac{V_{o(s)}}{V_{in(s)}} = \dfrac{T_o}{1 + \dfrac{s}{\omega_o}}$	$F_{(s)} = \dfrac{V_{o(s)}}{V_{in(s)}} = \dfrac{T_o s}{s + \omega_o}$
Sinusoidal response function	$F_{(j\omega)} = \dfrac{V_{o(j\omega)}}{V_{in(j\omega)}} = \dfrac{T_o}{1 + j\dfrac{\omega}{\omega_o}}$	$F_{(j\omega)} = \dfrac{V_{o(j\omega)}}{V_{in(j\omega)}} = \dfrac{T_o}{1 - j\dfrac{\omega_o}{\omega}}$
Magnitude of sinusoidal response	$\left\lvert\dfrac{V_{o(j\omega)}}{V_{in(j\omega)}}\right\rvert = \dfrac{T_o}{\sqrt{\left[1 + \left(\dfrac{\omega}{\omega_o}\right)^2\right]}}$	$\left\lvert\dfrac{V_{o(j\omega)}}{V_{in(j\omega)}}\right\rvert = \dfrac{T_o}{\sqrt{\left[1 + \left(\dfrac{\omega_o}{\omega}\right)^2\right]}}$
Phase of sinusoidal response	$\theta = -\tan^{-1}\left(\dfrac{\omega}{\omega_o}\right)$	$\theta = +\tan^{-1}\left(\dfrac{\omega_o}{\omega}\right)$

The values of the constants T_o, ω_o in the response equations are the parameters which distinguish one first-order function from another; they are determined by the component values used in the practical filter circuit. At a frequency $f_o = \omega_o/2\pi$ the magnitude of the sinusoidal response of a first-order filter is 3 dB down on the values of T_o and the filter produces a phase shift of 45°. It is usual to refer to f_o as the filter cut-off frequency and to regard this frequency as the one separating the pass band from the stop band of the filter. At frequencies in the stop band the slope of the magnitude response plot (dB–log f plot) of a first-order filter approaches a value of 20 dB/decade and the phase shift approaches a value of 90°. Low-pass filters produce phase lag (output lags on input), while high-pass filters produce phase lead (output leads on input). (First-order low-pass magnitude and phase plots are given in page 30 of the companion text).

Second-order filter functions

	Second-order low-pass	*Second-order high-pass*
Transfer function	$T_{(s)} = \dfrac{V_{o(s)}}{V_{in(s)}} = \dfrac{T_o\omega_o^2}{s^2 + 2\xi\omega_o s + \omega_o^2}$	$T_{(s)} = \dfrac{V_{o(s)}}{V_{in(s)}} = \dfrac{T_o s^2}{s^2 + 2\xi\omega_o s + \omega_o^2}$
Sinusoidal response function	$T_{(j\omega)} = \dfrac{V_{o(j\omega)}}{V_{in(j\omega)}} = \dfrac{T_o}{1 + j2\xi\dfrac{\omega}{\omega_o} - \dfrac{\omega^2}{\omega_o^2}}$	$T_{(j\omega)} = \dfrac{V_{o(j\omega)}}{V_{in(j\omega)}} = \dfrac{T_o}{1 - j2\pi\xi\dfrac{\omega_o}{\omega} - \dfrac{\omega_o^2}{\omega^2}}$
Magnitude of sinusoidal response	$\lvert T_{(j\omega)}\rvert = \dfrac{T_o}{\sqrt{\left[\left(1 - \dfrac{\omega^2}{\omega_o^2}\right)^2 + 4\xi^2\dfrac{\omega^2}{\omega_o^2}\right]}}$	$\lvert T_{(j\omega)}\rvert = \dfrac{T_o}{\sqrt{\left[\left(1 - \dfrac{\omega_o^2}{\omega^2}\right)^2 + 4\xi^2\dfrac{\omega_o^2}{\omega^2}\right]}}$

Phase of sinusoidal response $\theta = -\tan^{-1}\left(\dfrac{2\xi\omega/\omega_o}{1-\omega^2/\omega_o^2}\right)$ $\theta = +\tan^{-1}\left(\dfrac{2\xi\omega_o/\omega}{1-\omega_o^2/\omega^2}\right)$

Second-order band-pass

Transfer function $$T_{(s)} = \frac{V_{o(s)}}{V_{in(s)}} = \frac{T_o 2\xi\omega_o s}{s^2 + 2\xi\omega_o s + \omega_o^2}$$

Sinusoidal response function $$T_{(j\omega)} = \frac{V_{o(j\omega)}}{V_{in(j\omega)}} = \frac{T_o}{1 + \dfrac{1}{2\xi} j\left(\dfrac{\omega}{\omega_o} - \dfrac{\omega_o}{\omega}\right)}$$

Magnitude of sinusoidal response $$|T_{(j\omega)}| = \frac{T_o}{\sqrt{\left[1 + \dfrac{1}{4\xi^2}\left(\dfrac{\omega}{\omega_o} - \dfrac{\omega_o}{\omega}\right)^2\right]}}$$

Phase of sinusoidal response $$\theta = -\tan^{-1}\left[\frac{1}{2\xi}\left(\frac{\omega}{\omega_o} - \frac{\omega_o}{\omega}\right)\right]$$

T_o, the pass-band gain, $f_o = \omega_o/2\pi$, the natural frequency, and ξ, the damping factor, are the distinguishing parameters associated with a second-order response. The values of the parameters are determined by the components used in the practical circuit realization of the response.

Second-order low-pass and band-pass functions are characterized by a magnitude response which exhibits an eventual 40 dB/decade roll-off in the stop band. Second-order band-pass functions have a magnitude response which eventually rolls off at 20 dB/decade either side of f_o at frequencies well removed from f_o.

The value of the damping factor of a second-order function governs the shape of its magnitude and phase response plot at frequencies in the neighbourhood of f_o. As the value of ξ is reduced below $\xi = 1/\sqrt{2}$, the magnitude response shows increasing amounts of peaking in the region of f_o (equations relating the gain peaking and transient response of a second-order low-pass system to its damping factor are given in Section 3.8.2 of the companion text). Plots showing

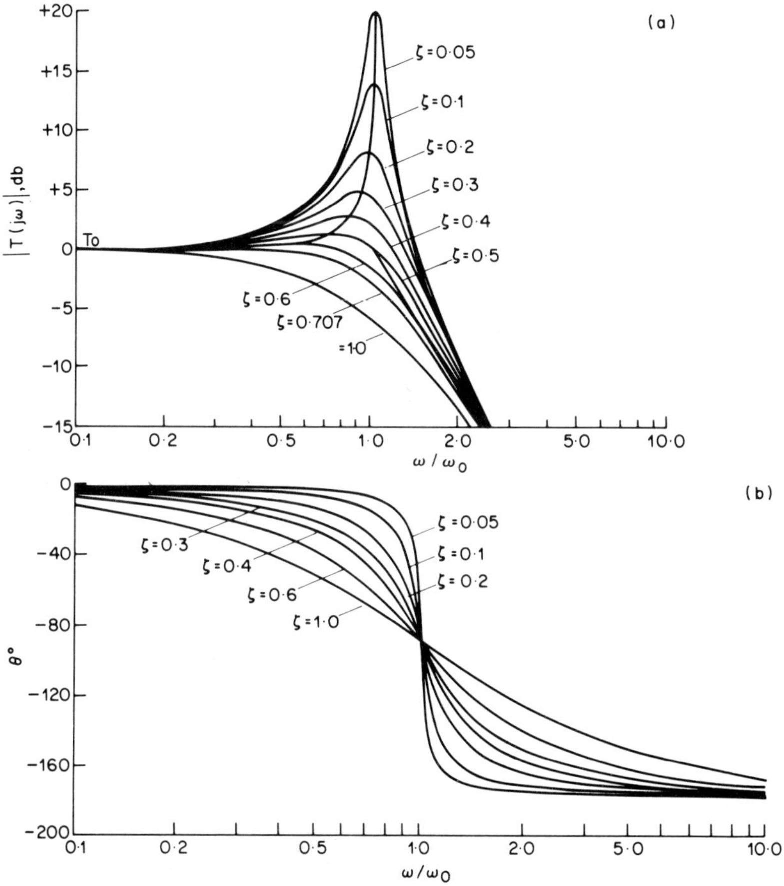

Figure 32 Magnitude and phase response of a second-order low-pass filter: (a) magnitude versus frequency; (b) phase versus frequency

the shape of the magnitude and phase response of a second-order low-pass system are given in Figure 32. Note that the phase shift is 90° at $f = f_o$ and that it approaches a limiting value of 180° in the stop band. Phase lag occurs for a low-pass system, and phase lead for a high-pass system.

The attenuation characteristics associated with a second-order band-pass characteristic are shown in Figure 33. It is usual to speak

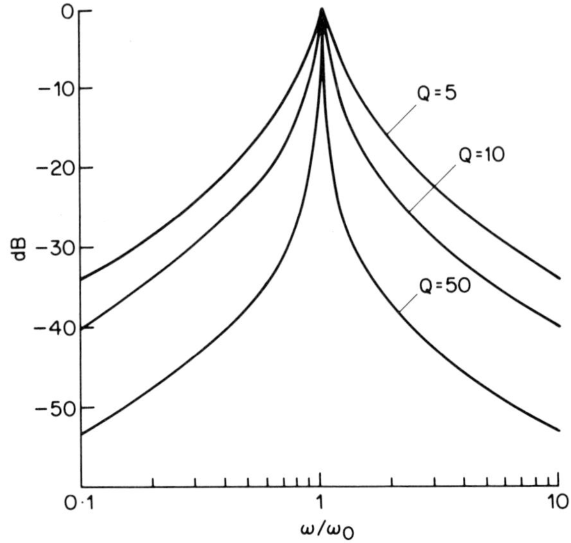

Figure 33 Magnitude versus frequency for a second-order band-pass characteristic

of the Q factor of a band-pass characteristic. The Q factor is related to the damping factor by $Q = 1/2\xi$. The higher the Q factor, the more peaked is the magnitude response. At a frequency $f = f_o = \omega_o/2\pi$ it has a value T_o. Q is related to attenuation characteristics by

$$Q = f_o/(f_2 - f_1) \qquad (\sqrt{f_1 f_2} = f_o)$$

f_2 and f_1 are the frequencies either side of f_o at which the magnitude of the response (the gain) is 3 dB down on the value T_o. $f_2 - f_1$ defines the bandwidth of the filter characteristic. The phase shift associated with a second-order band-pass characteristic is $+45°$ at $f = f_1$, zero at $f = f_o$ and $-45°$ at $f = f_2$.

The reader is asked to verify the above general points relating to filter functions by the measurements and observations which he makes on active filter circuits connected up on the experimental circuit board.

Experimental procedure for active filter experiments

The same general experimental procedure is suitable for measuring the sinusoidal response of all the active filter circuits given:

1. Patch the given filter circuit on the experimental circuit board.
2. Apply a sinusoidal signal as an input test signal. Monitor the input and output signals with a dual-trace oscilloscope.
3. Vary the frequency of the test signal and take readings which allow you to plot the magnitude and phase characteristics of the filter.

Graphs are conveniently plotted on a piece of three-decade log–linear graph paper. The frequency is plotted on the horizontal log scale, while magnitude (dB) and phase θ are plotted on the linear vertical scale.

The phase relationship between input and output signals can be estimated from the observed oscilloscope traces. The magnitude of the filter response at a particular frequency is determined as the ratio (output signal amplitude)/(input signal amplitude) expressed in dB. Measurements are facilitated by using a high-impedance a.c. voltmeter with a dB scale (as in Experiment 4). The magnitude of the response in dB is then obtained as: (output reading in dB) – (input reading in dB).

The amplitude of the input test signal is not of critical importance but it should not be so large as to cause the operational amplifier output limits to be reached.

In each filter experiment, before recording readings, it is a useful practice to make a quick sweep-through of the test signal frequency. In this way the general trend of the filter response may be observed and an approximate value of the filter's f_o can be established. Readings may then be confined to frequencies a decade or so on either side of f_o.

First-order active low-pass filter

Connections for a first-order active low-pass filter are given in

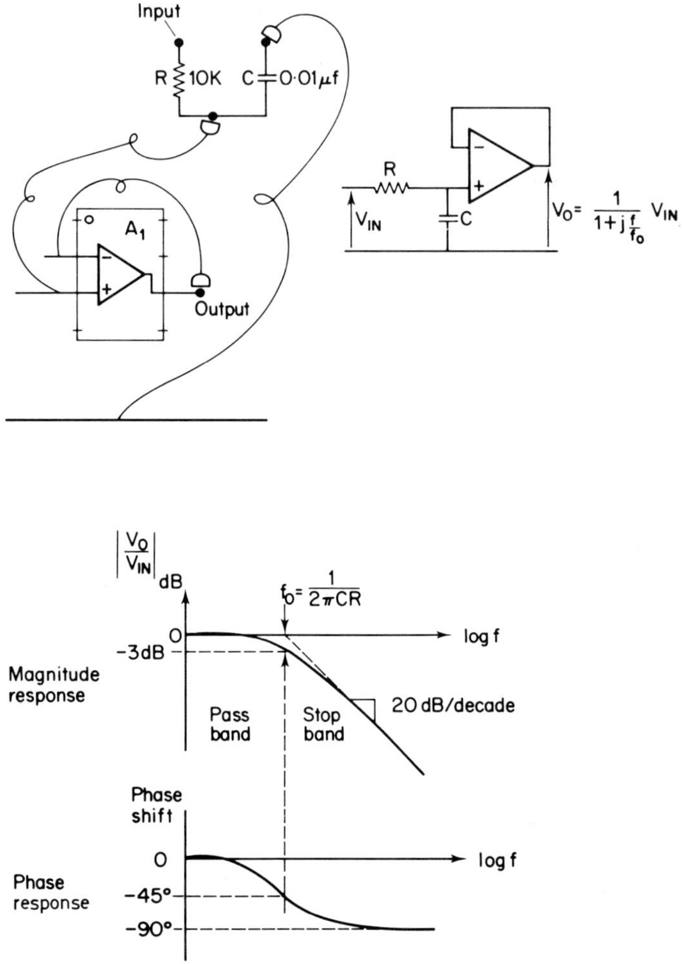

Figure 34 A first-order low-pass active filter

Figure 4. The operational amplifier acts simply as a unity-gain output buffer; it prevents the output loading errors which are associated with passive *CR* filter networks. The transfer function parameters are $T_o = 1, f_o = \omega_o/2\pi = 1/2\pi CR$. A pass-band gain T_o other than unity is easily obtained by connecting the operational amplifier as a follower with gain (see Experiment 1).

Measure the sinusoidal response (magnitude and phase) of the filter. Note that at frequency $f = f_o$ the magnitude of the response is 3 dB down on T_o and the phase shift is $-45°$.

Observe the response of the filter to a square-wave input test signal. Measure the output rise time T_r as in Experiment 4. Note that $T_r \approx 1/3f_o$.

First-order active high-pass filter

Connections for a first-order active high-pass filter are given in Figure 35. The operational amplifier acts simply as an output buffer.

Measure the sinusoidal response of the filter; relate your observations to the first-order equations given previously.

Note that the response of all active high-pass filters exhibits a roll-off at high frequencies. This is caused by the roll-off in the open-loop gain frequency response of the operational amplifier. Open-loop gain roll-off causes a reduction in loop gain. At frequencies for which the magnitude of the loop gain is less than unity the frequency response of the active high-pass filter follows the open-loop frequency response of the amplifier.

Second-order active low-pass filters

There are many different operational amplifier circuit configurations for the realization of second-order filter functions 1–3. One of the simplest methods, which is quite suitable for low-Q filter applications, is based upon the use of an operational amplifier as a so-called 'voltage-controlled voltage source' (VCVS). An ideal VCVS is a voltage amplifier with infinite input impedance, zero output impedance and a defined voltage gain. An operational amplifier, arranged as a voltage follower, provides a close practical approximation of an ideal VCVS. VCVS filters were first described by Sallen and Key[4] and are sometimes referred to as 'Sallen–Key filters'.

A general circuit for a second-order low-pass Sallen–Key filter is

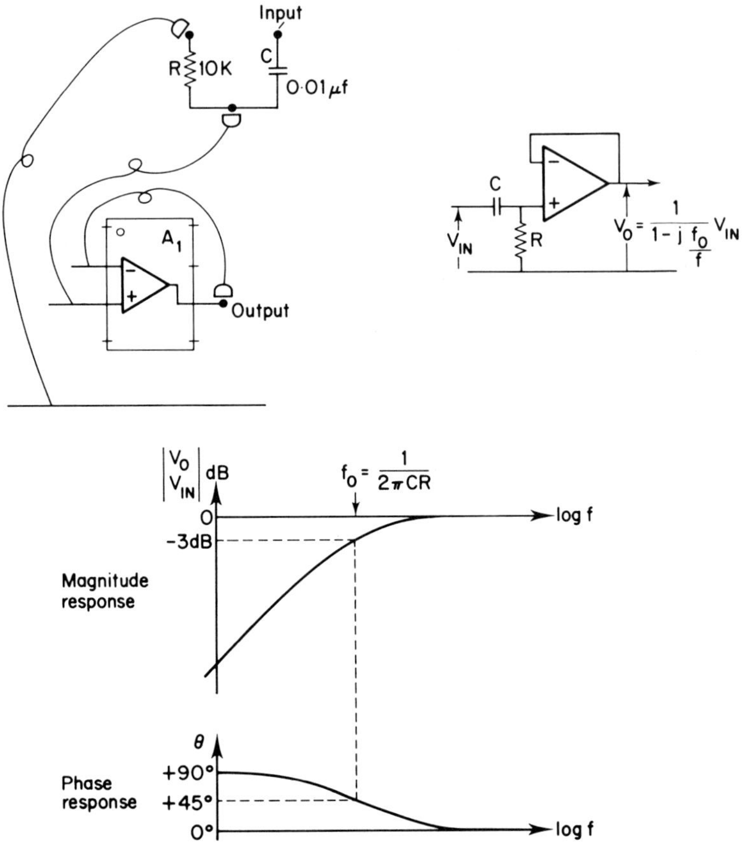

Figure 35 A first-order high-pass active filter

shown in Figure 36a. The design equations which relate component values to the second-order transfer function parameters are

$$T_o = K = 1 + \frac{R_4}{R_3} \qquad 2\pi f_o = \omega_o = \sqrt{\frac{1}{R_1C_1R_2C_2}}$$

$$\frac{1}{Q} = 2\xi = \sqrt{\frac{R_2C_2}{R_1C_1}} + \sqrt{\frac{R_1C_2}{R_2C_1}} + (1-K)\sqrt{\frac{R_1C_1}{R_2C_2}}$$

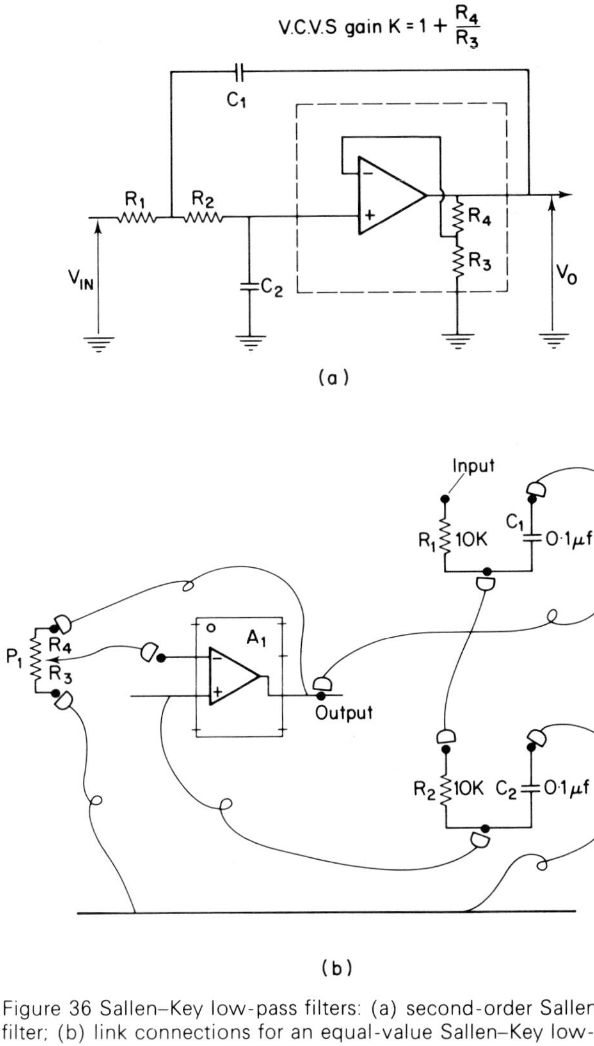

Figure 36 Sallen–Key low-pass filters: (a) second-order Sallen–Key low-pass filter; (b) link connections for an equal-value Sallen–Key low-pass filter

Clearly, particular values for f_o and ξ can be fixed by many different sets of component values. In order to facilitate the choice of component values it is usual to make some simplifying design assumptions.

One approach is to make $R_1 = R_2 = R$, $C_1 = C_2 = C$; this is referred to as an equal-value Sallen–Key filter. This makes

$$f_o = \frac{\omega_o}{2} = \frac{1}{2\pi CR} \qquad \frac{1}{Q} = 2\xi = 3 - K \qquad T_o = K = 1 + \frac{R_4}{R_3}$$

The damping factor can be adjusted to a required value by adjusting K (adjusting the operational amplifier gain-setting resistors). Varying K does not change the natural frequency f_o of the filter.

The link connections which are required to patch an equal-value Sallen–Key filter on the experimental circuit board are shown in Figure 36b. Measure the sinusoidal response of the filter for different values of the damping factor ξ. Relate your observations to the second-order equations and discussion given previously.

A damping factor (ξ) value is conveniently established by experimentally setting its required K value; adjust potentiometer P_1 to give the follower-connected operational amplifier its required gain.

A second-order Butterworth filter characteristic is obtained with $2\xi = \sqrt{2}$, a K setting of $K = 3 - 2\xi = 1.59$.

A second-order Tchebyscheff characteristic is obtained with $2\xi = 1.045$, a K setting of $K = 3 - 2\xi = 1.96$.

A second-order Bessel characteristic is obtained with $2\xi = 1.73$, a K setting of $K = 3 - 2\xi = 1.27$.

The names 'Butterworth', 'Tchebyscheff', and 'Bessel' refer to filter functions which are devised to optimize specific aspects of filter response. An ideal filter would pass unchanged all signals with frequencies in its pass band and completely reject signal frequencies in its stop band; real filters do not give such a sharply defined transition between pass and stop bands. The final rate of attenuation in the stop band of a filter is governed by the order of the filter: first order, 20 dB/decade; second order, 40 dB/decade; third order 60 dB/decade; and so on. However, in second- and higher-order filters the nature of the response in the transition from pass band to stop band is governed by factors other than the order of the filter. In second-order filters it is governed by the damping factor associated with the filter transfer function. In higher-order filters, obtained by cascading first- and second-order filters, it is governed by the f_0 and ξ values of the filters in the cascade. The design of third- and higher-order filters involves the setting of the f_0 and ξ values of the individual filter stages in the cascade in order to achieve a desired overall response 1–3.

Butterworth and Tchebyscheff filter functions are approximations of an ideal magnitude–frequency characteristic. A Butterworth design gives a magnitude–frequency characteristic which is said to be maximally flat in the pass band. A Tchebyscheff design has a sharper cut-off characteristic than a Butterworth design but its pass band is not completely flat.

Butterworth and Tchebyscheff designs are not concerned with phase response, but phase relationships are important in determining the effect that a filter has on an input signal with a complex non-sinusoidal waveform. It is desirable, in order to minimize phase distortion of a complex wave, that the phase shift θ produced by a filter should vary linearly with frequency. A non-linear phase–frequency relationship can cause the transient response of a low-pass filter to exhibit pronounced overshoot and ringing.

A Bessel filter design is an approximation of an ideal linear phase–frequency characteristic but its magnitude–frequency characteristic is a much poorer approximation of the ideal than either the Butter-

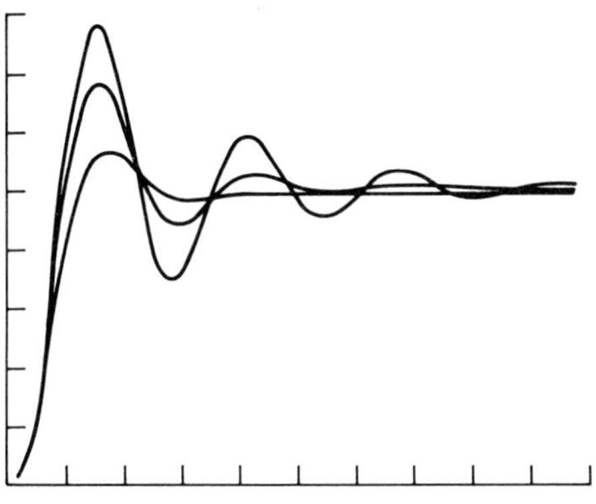

Figure 37 The transient response of a second-order low-pass filter for different values of damping factor

worth or the Tchebyscheff designs. A Bessel design gives a filter with the best transient response.

Measure the transient response of the test circuit of Figure 36b. Apply a low-frequency ($f < f_o$) square wave as an input test signal and observe the output waveform produced by the filter. Results obtained in a test of this kind are illustrated by the oscilloscope traces given in Figure 37. As the damping factor is reduced the output shows increasing amounts of overshoot and ringing. In obtaining the traces shown in Figure 37 the oscilloscope vertical sensitivity was adjusted to give equal output step size for the different K values.

Sallen–Key low-pass filter with $K = 1$

An alternative design procedure for a Sallen–Key low-pass filter is as follows. Make $K = 1$ by connecting the operational amplifier as a unity-gain follower. Make $R_1 = R_2 = R$, giving

$$f_o = \frac{\omega_o}{2\pi} = \frac{1}{2\pi R\sqrt{C_1 C_2}} \text{ and } \frac{1}{Q} = 2\xi = 2\sqrt{\frac{C_2}{C_1}}$$

A Butterworth response with $f_o = 1/2\pi\sqrt{2}CR$ and $2\xi = \sqrt{2}$ is obtained by choosing $C_2 = C$ and $C_1 = 2C$. Link connections for such a filter design are shown in Figure 38.

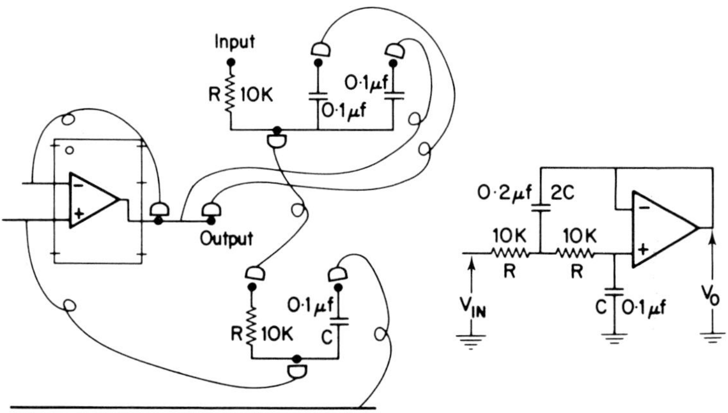

Figure 38 A Sallen–Key low-pass filter with $K = 1$, Butterworth response

Second-order Sallen–Key high-pass filter

A second-order Sallen–Key high-pass filter is shown in Figure 39a. The high-pass circuit is obtained by transposing the *R*'s and *C*'s in the low-pass Sallen–Key circuit. The design equations which relate

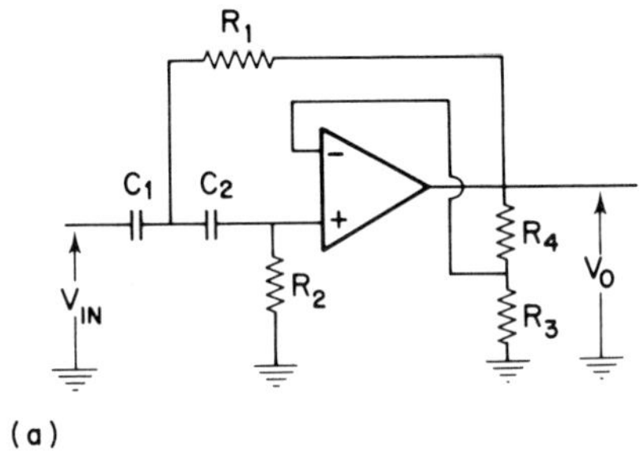

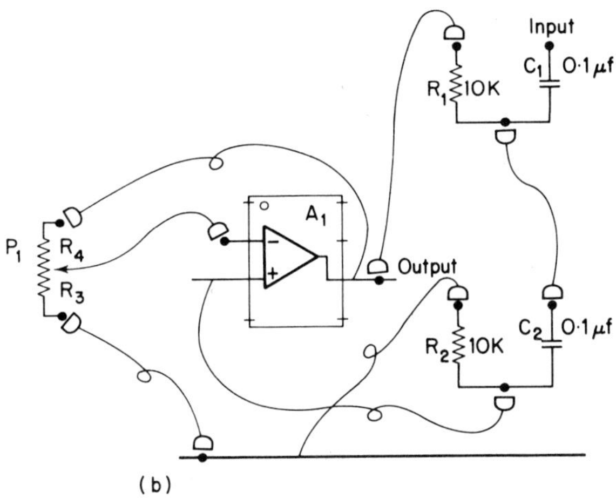

Figure 39 Sallen–Key high-pass filters: (a) a Sallen–Key high-pass filter; (b) link connections for an equal-value Sallen–Key high-pass filter

component values to the parameters of the general second-order high-pass transfer function are

$$2\pi f_o = \omega_o = \sqrt{\frac{1}{R_1 C_1 R_2 C_2}} \qquad T_o = K = 1 + \frac{R_4}{R_3}$$

$$\frac{1}{Q} = 2\xi = \sqrt{\frac{R_1 C_1}{R_2 C_2}} + \sqrt{\frac{R_1 C_2}{R_2 C_1}} + (1 - K)\sqrt{\frac{R_2 C_2}{R_1 C_1}}$$

As before, it is convenient to make design assumptions in order to simplify the choice of components for a particular response. Making $C_1 = C_2 = C$ and $R_1 = R_2 = R$ gives an equal-value Sallen–Key high-pass filter with the design equation simplified to $2\pi f_o = \omega_o = 1/RC$, $1/Q = 2\xi = 3 - K$. Q may be adjusted by varying the K value without influencing f_o. Link connections for the equal-value Sallen–Key high-pass filter are shown in Figure 39b.

Measure the sinusoidal response of the high-pass filter of Figure 39b for different values of damping factor. Relate your observations to the second-order high-pass equations and discussion given previously.

An alternative design assumption is to make $K = 1$ and $C_1 = C_2 = C$. This makes

$$f_o = \frac{1}{2\pi C\sqrt{R_1 R_2}} \quad \text{and} \quad \frac{1}{Q} = 2\xi = 2\sqrt{\frac{R_1}{R_2}}$$

A Butterworth response with $2\xi = \sqrt{2}$ is obtained by choosing $R_1 = R$ and $R_2 = 2R$, giving $f_o = 1/2\pi\sqrt{2}CR$. Readers are asked to devise link connections on the experimental board which give such a design configuration.

Second-order band-pass filters

A second-order band-pass active filter based upon the use of an operational amplifier as a VCVS is shown in Figure 40. The design equations relating component values to the parameters of the second-order band-pass transfer function are rather cumbersome in the general case[2]. In the equal-value case, $C_1 = C_2 = C$, $R_1 R_2 = R_3 = R$, they simply to

$$\omega_o = 2\pi f_o = \frac{\sqrt{2}}{CR} \qquad T_o = \frac{K}{5-K} \qquad \frac{1}{Q} = 2\xi = \frac{5-K}{\sqrt{2}}$$

Link connections for an equal-value VCVS band-pass filter are shown in Figure 40b. Measure the magnitude and phase of the sinusoidal response of the filter for various Q values. Relate your observations to the general second-order band-pass filter behaviour. A particular Q value is established by setting its appropriate K

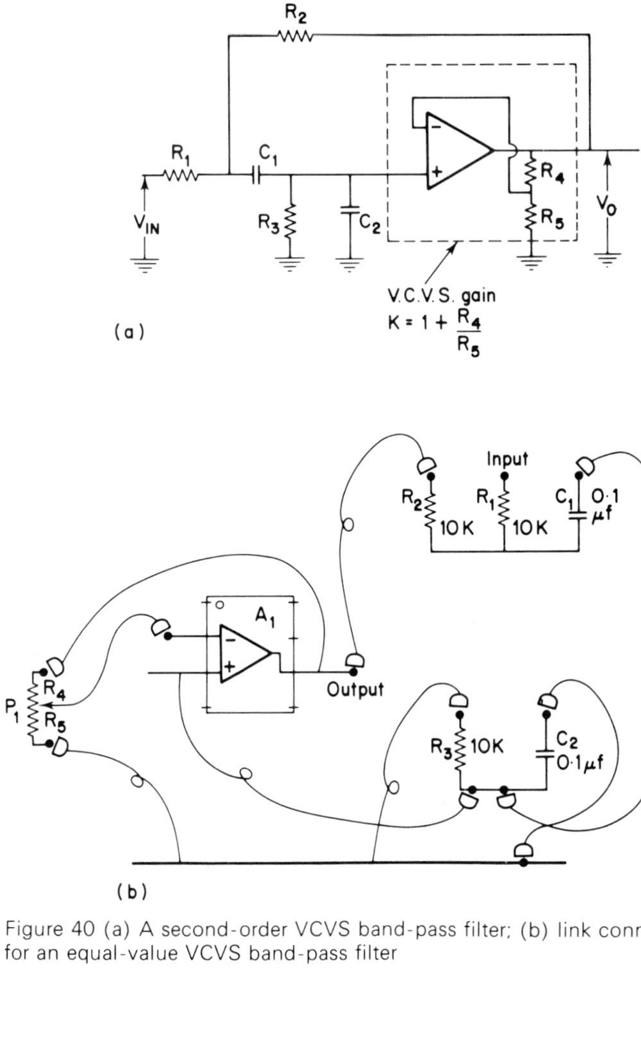

Figure 40 (a) A second-order VCVS band-pass filter; (b) link connections for an equal-value VCVS band-pass filter

value. For example $Q = 5$ requires $1/5 = (5 - K)/\sqrt{2}$, i.e. $K = 5 - \sqrt{2}/5 = 4.72$. Remember that K represents the gain of the follower-connected operational amplifier and its value can be set experimentally by adjustment of the gain-setting potentiometer P_1.

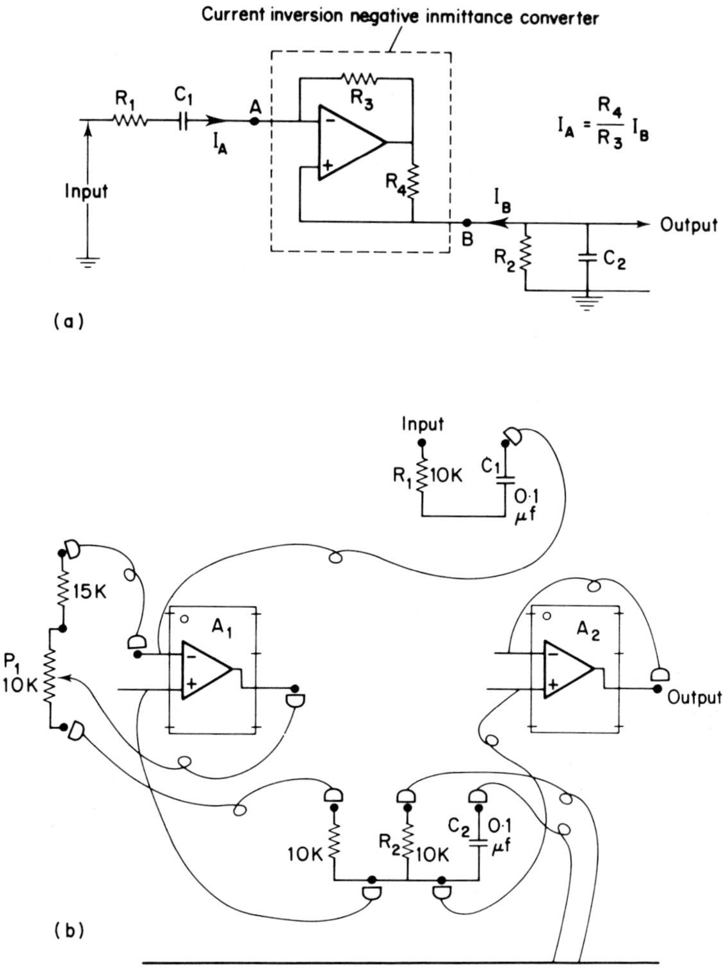

Figure 41 (a) An INIC band-pass filter; (b) link connections for an INIC band-pass filter and output buffer

The VCVS band-pass filter configuration is not suitable for use with high Q values ($Q > 10$). High-Q VCVS filters have transfer function parameters which are very sensitive to component values.

A second-order active band-pass filter configuration which is not as sensitive to component values as the VCVS configuration is given in Figure 41. The configuration makes use of an operational amplifier acting as a current inverter (see page 151 of companion text) and is referred to as an 'INIC band-pass filter'. 'INIC' stands for 'ideal negative immittance converter'[1,2]. The operational amplifier maintains its two input terminals at the same potential. If a current I_A enters at point A the amplifier causes a current $I_B = (R_3/R_4)I_A = GI_A$ to flow towards the point B. If an impedance Z_B is connected between point B and earth, the operational amplifier circuit presents an input impedance $Z_A = -GZ_B$ between point A and earth.

The output of an INIC active filter is not, as in the case of the other active filter circuits considered, available at the low-impedance output terminal of an operational amplifier. Practical INIC filters need to be used together with an output buffer stage to prevent loading errors.

The design equations which relate the component values in the circuit of Figure 41a to the parameters of the general second-order band-pass transfer function are

$$T_o = \frac{G}{C_2/C_1 + R_1/R_2 - G}$$

where

$$G = R_3/R_4$$

$$\omega_o = 2\pi f_o = 1/\sqrt{R_1C_1R_2C_2}$$

$$\frac{1}{Q} = 2\xi = \sqrt{\frac{R_2C_2}{R_1C_1}} + \sqrt{\frac{R_1C_1}{R_2C_2}} - G\sqrt{\frac{R_2C_1}{R_1C_2}}$$

Component selection for a particular response is facilitated by setting $C_1 = C_2 = C$ and $R_1 = R_2 = R$. The design equations simplify to

$$T_o = \frac{G}{2-G} \qquad \omega_o = 2\pi f_o = \frac{1}{CR} \qquad \frac{1}{Q} = 2\xi = 2 - G$$

Link connections for such an arrangement are given in Figure 41b.

A second operational amplifier is shown connected as an output buffer. Measure the sinusoidal response of the filter circuit for various values of Q. Relate your observations to the general second-order band-pass response.

Different values for Q may be established by adjusting the setting of potentiometer P_1. Adjustment of P_1 alters the ratio R_3/R_4 which determines the 'gain' G of the current inverter. For example, since $2 - G = 1/Q$, a Q value of 20 requires that G be set to a value 1.95; with $G = 1.95$ the centre frequency gain $T_o = G/(2 - G) = 1.95/0.05 = 39$. Thus adjustment of P_1 to define a measured centre frequency gain of 39 in effect establishes a Q value of 20.

Second-order band-reject filter (notch filter)

Band-reject filters have properties which are closely related to those of band-pass filters. A general expression for the transfer function of a second-order band-reject filter may be written as

$$\underset{\text{B.Reject}}{T_{(s)}} = T_o \frac{s^2 + \omega_o^2}{s^2 + 2\xi\omega_o s + \omega_o^2}$$

An expression for the steady–state sinusoidal response is obtained by substituting $s = j\omega$ which after rearrangement yields

$$\underset{\text{B.Reject}}{T_{(j\omega)}} = T_o \frac{1}{1 + \frac{j}{Q}\left(\frac{\omega_o \omega}{\omega_o^2 - \omega^2}\right)}$$

The magnitude of the sinusoidal response is

$$\left| \underset{\text{B.Reject}}{T_{(j\omega)}} \right| = \frac{1}{\sqrt{\left[1 + \frac{1}{Q^2}\left(\frac{\omega_o \omega}{\omega_o^2 - \omega^2}\right)^2\right]}}$$

The phase shift θ is

$$\theta = \tan^{-1}\left[\frac{1}{Q}\left(\frac{\omega_o \omega}{\omega_o^2 - \omega^2}\right)\right]$$

We may write

$$Q = \frac{\omega_o}{\omega_2 - \omega_1} \qquad \omega_o^2 = \omega_2\omega_1$$

where ω_2 and ω_1 are the frequencies at which the magnitude of the response is 3 dB down on T_o.

Examination of the general band-pass transfer function given previously shows that we may write

$$\underset{\text{B.Reject}}{T_{(s)}} = T_o - \underset{\text{B.Pass}}{T_{(s)}}$$

A band-reject filter can be realized by summing the output of an inverting band-pass filter with its suitably scaled input signal (see pages 313 and 314 of companion text). The patching of such a filter requires two experimental boards. Readers interested in pursuing this are referred to the circuit given in Figure 8.14 of the companion text.

An alternative band-reject active filter configuration which can be connected up on a single experimental board is shown in Figure 42a. The circuit is based upon the use of a modified twin-T network[5], and it has provision for varying the Q factor of the response. With the components proportioned as indicated in Figure 42a the design equations which relate component values to the parameters of the general second-order band-reject transfer function are

$$T_o = 1 \qquad \omega_o = 1/CR \qquad 1/Q = 2\xi = 4(1 - m)$$

Link connections for an experimental version of the notch filter are given in Figure 42b. Measure the sinusoidal response (magnitude and phase) of the filter circuit for various Q values. A particular Q value is obtained by adjusting potentiometer P_1 to give the necessary attenuation factor m:

$m = 0$ gives $1/Q = 4 \times 1$ i.e. $Q = \frac{1}{4}$

$m = 0.75$ gives $1/Q = 4 \times 0.25$ i.e. $Q = 1$

$m = 1$ theoretically gives an infinite Q value but in practice the observed response will depend very much upon component matching. It is inadvisable to use very-high-Q notch filters because of their high sensitivity to component values. Q values may be measured using the relationship

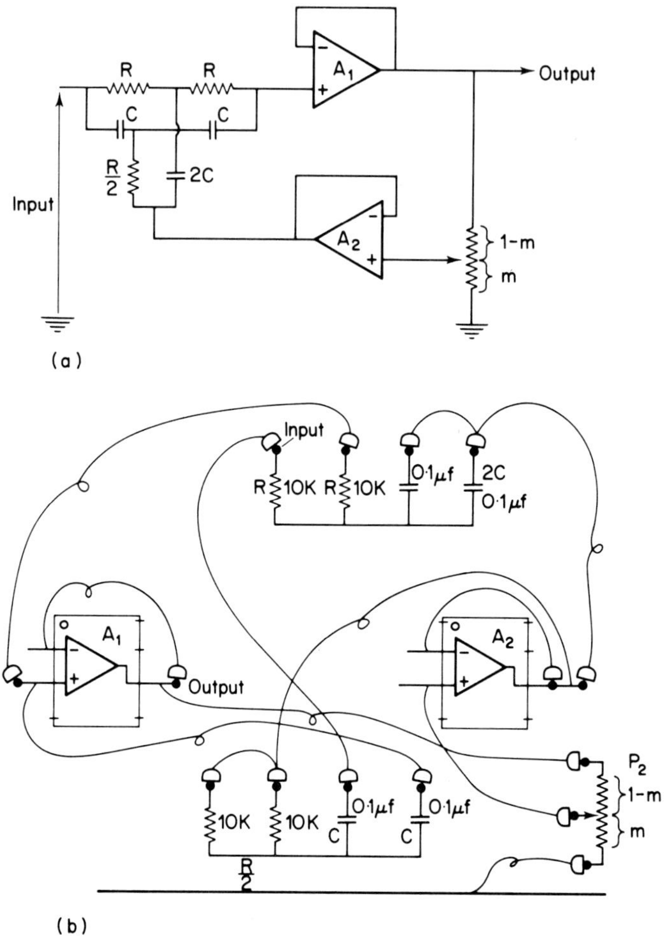

Figure 42 A band-reject filter (notch filter): (a) active twin-T band-reject filter with variable *Q*; (b) link connections for twin-T active band-reject filter

$$Q = f_o/(f_2 - f_1)$$

f_o is the centre frequency of the notch at which the filter produces a theoretically infinite attenuation. The depth of the notch which is obtained in practice depends upon component matching. f_2 and f_1 are the upper and lower frequencies at which the filter produces an attenuation of 3 dB.

Note that if the test signal waveform contains harmonic distortion the harmonics will not be rejected by the notch filter when the test signal frequency is set at f_o.

General points relating to active filters

The range of response parameter values for the active filters which can be assembled on the experimental board is clearly limited by the values of the components available on the board. Readers wishing to employ one or other of the filter designs given in a practical filtering application will need to choose component values in order to define a specific response.

A particular *CR* product can obviously be set by a range of different *C*'s and *R*'s. Stable close-tolerance capacitors of low value are available in a wider range of values and are less expensive and physically smaller than large-value capacitors. Hence, when choosing component values for a low-frequency filter it is advantageous to choose a small value of *C* and a large value of *R*. The upper limit to resistor magnitude is governed by offset error due to amplifier bias current (see Experiment 2). By using low-bias-current FET input operational amplifiers, offset errors can be kept small even though resistor values are large. For example, a TL081 BiFet amplifier is specified as having a worst-case bias current of 100 pA at 25 °C; this means that with a source resistance as high as 10 MΩ the input offset error due to bias current is still only 1 mV.

The stability of filter response parameters is directly influenced by the stability of the passive components of the filter. Stable close-tolerance resistors are available in a wider range of values than capacitors; hence when setting a *CR* product it is advisable first to choose a value for *C* and then to calculate the value required for *R*. Low-temperature-coefficient resistors, (say metal film) should be used for stability; they can be trimmed or padded to an exact value with lower-cost carbon film resistors. The greater temperature coefficient of resistance of the carbon film resistor has only a small effect on the overall temperature coefficient.

High-*Q* band-pass filters are particularly sensitive to component tolerance and stability. A single very-high-*Q* filter can become unstable and produce oscillations because of changes in its component values. If Q values greater than say 50 are required it is better to obtain such a response by cascading lower-Q stages.

The equations given for the active filter transfer functions are based upon the assumption that there is sufficient gain in the feedback loop (a large enough value of the loop gain $\beta A_{OL_{(jf)}}$ to make the errors due to the finite open-loop gain of the amplifier negligible. Sinusoidal transfer functions must be multiplied by the gain error factor $1/1 + 1/\beta A_{OL_{(jf)}}$ in order to take account of the finite open-loop gain of the operational amplifier (see Section 2.3.1 of companion text). In the case of an active band-pass filter the product of the filter's centre frequency and gain must be less than one-tenth of the operational amplifier's unity-gain bandwidth product to ensure that the magnitude of the loop gain at the centre frequency is greater than 10.

References

1 Huelsman, L.P., *Theory and Design of Active R.C. Networks*, McGraw Hill, (1968).
2 Tobey, G. & Huelsman, L.P., *Operational Amplifiers Design and Applications*, McGraw Hill, (1971).
3 Clayton, G. B., *Linear Integrated Circuit Applications*, MacMillan, (1975).
4 Nicholson, P.F., *The BiFet Design Manual*, Texas Instruments Ltd, (1980).
5 Farrer, T.E., *Elect. Engg.*, **139**, No. 470, 219.

Exercises 11

1 A sinusoidal input signal of amplitude 5 V is applied to the first-order low-pass filter circuit in Figure 34 ($C = 0.01\ \mu F$, $R = 10\ k\Omega$). What will be the amplitude of the output signal and its phase relative to the input signal when the frequency of the input signal is (a) 100 Hz, (b) 159 Hz, (c) 1.59 kHz, (d) 100 kHz?

2 Find the attenuation and phase shift produced by a second-order low-pass filter at frequencies 50, 100, 125 and 200 Hz if the filter has pass-band gain $T_o = 1$, natural frequency $f_o = 100$ Hz and damping factor ξ (a) $1/\sqrt{2}$ (Butterworth response), (b) 0.86 (Bessel response), and (c) 0.523 (Tchebyscheff response).

3 Choose component values for use in an equal-value Sallen–Key low-pass filter in order to give it the f_o and ξ values given in the previous exercise (use $C = 0.002\ \mu F$). What pass-band gains will these filters have?

4 A second-order band-pass filter response has $Q = 20$ and natural frequency $f_o = 200$ Hz.
(a) At what frequencies will the filter produce an attenuation of 3 dB on the centre frequency gain and what will the phase shift be at these frequencies?
(b) Suggest component values for use in an equal-value VCVS realization of the above response (Figure 40, use $C = 0.002\ \mu F$). What will the centre frequency gain of this filter be?
(c) Suggest component values for use in an equal-value INIC realization of the above response (Figure 41a, use $C = 0.002\ \mu F$). What centre frequency gain will this filter have?

Experiment 12
Comparators

A comparator is a functional unit which has two input lines and a single output. It is used to indicate which is the larger of the signals applied to its inputs. The output of a comparator can take on only one or the other of two voltage levels, a high level and a low level. The functional operation performed by an ideal comparator is such that its output changes state whenever the signals applied to its two inputs pass through equality with one another. In a sense, a comparator compares its two input signals and indicates by the state of its output which of the two signals is the bigger.

Experiments to investigate the use of an operational amplifier as a comparator are described in this section. Comparator applications of operational amplifiers are discussed in Section 7.1 of the companion text. The use of operational amplifiers as comparators is restricted to those applications not requiring very fast output switching between states. Applications requiring fast switching are best performed by integrated circuit devices specifically designed as comparators.

A simple comparator application

Make the link connection shown in Figure 43. Apply a low-frequency sinusoidal test signal (amplitude 5 V or so) directly to the non-inverting input terminal of the amplifier. Apply a variable d.c. voltage as a reference level to the inverting input terminal of the amplifier.

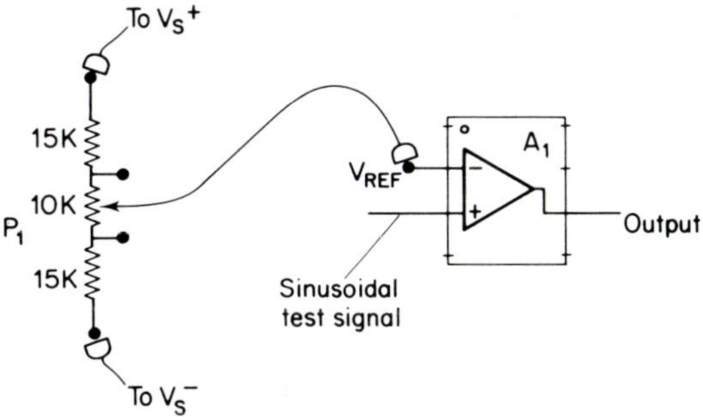

Figure 43 An open-loop operational amplifier performs a comparator function

Monitor the amplifier output and the sinusoidal test signal with a dual-trace oscilloscope.

The amplifier acts open loop; because of the very large open-loop gain its output switches between saturation limits every time the signals at the two inputs pass through approximate equality.

A regenerative comparator

Make the link connections shown in Figure 44. There is positive feedback applied to the amplifier via resistors R_1 and R_2; the circuit acts as a regenerative comparator.

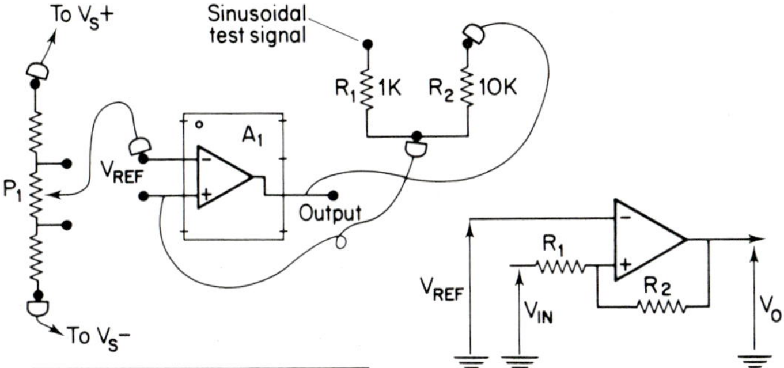

Figure 44 An operational amplifier with positive feedback applied to it acts as a regenerative comparator

Apply a low-frequency sinusoidal input test signal (amplitude say 8 V or so), and monitor the amplifier output and input test signal with a dual-trace oscilloscope. Verify that the input threshold levels at which the output switches between states are determined by

$$\underset{+\,-}{V_{\text{th}}}\frac{R_2}{R_1+R_2}+\underset{+\,-}{V_{\text{o sat}}}\frac{R_1}{R_1+R_2}=V_{\text{ref}}$$

That is

$$\text{Upper threshold}\quad \underset{+}{V_{\text{th}}}=V_{\text{ref}}\frac{R_1+R_2}{R_2}+\left|\underset{-}{V_{\text{o sat}}}\right|\frac{R_1}{R_2}$$

$$\text{Lower threshold}\quad \underset{-}{V_{\text{th}}}=V_{\text{ref}}\frac{R_1+R_2}{R_2}-\left|\underset{+}{V_{\text{o sat}}}\right|\frac{R_1}{R_2}$$

A display of the comparator transfer curve (output–input) can be obtained by using the oscilloscope in the *x–y* mode (input test signal, horizontal deflection; output, vertical deflection).

An alternative configuration is obtained by interchanging the reference and input signal points. As before, output switching occurs when the voltage levels at the two input terminals of the amplifier pass through equality. In the alternative configuration the threshold levels of the input signal are

$$\underset{+}{V_{\text{th}}}=V_{\text{ref}}\frac{R_2}{R_1+R_2}+\left|\underset{+}{V_{\text{o sat}}}\right|\frac{R_1}{R_1+R_2}$$

$$V_{\text{th}}=V_{\text{ref}}\frac{R_2}{R_1+R_2}-\left|\underset{-}{V_{\text{o sat}}}\right|\frac{R_1}{R_1+R_2}$$

Exercises 12

1 A regenerative comparator is required to give a positive output transition when the input signal increases through the value +2 V and a negative output transition when the input decreases through the value −2 V. Suggest suitable component values for use in the circuit of Figure 44 if the operational amplifier has output saturation limits ±10 V.

2 A regenerative comparator is required which gives a negative

output transition when its input signal increases through the value +5 V and a positive output transition when its input signal decreases through the value +3 V. Suggest a suitable circuit arrangement for this purpose using an operational amplifier with output saturation limits ±10 V.

Experiment 13
A free-running multivibrator

In this experiment the student is asked to examine the action of an operational amplifier as a free-running multivibrator. Multivibrator circuits are discussed in Section 7.2 of the companion text.

Experimental procedure

Make the link connections shown in Figure 45. Observe the circuit

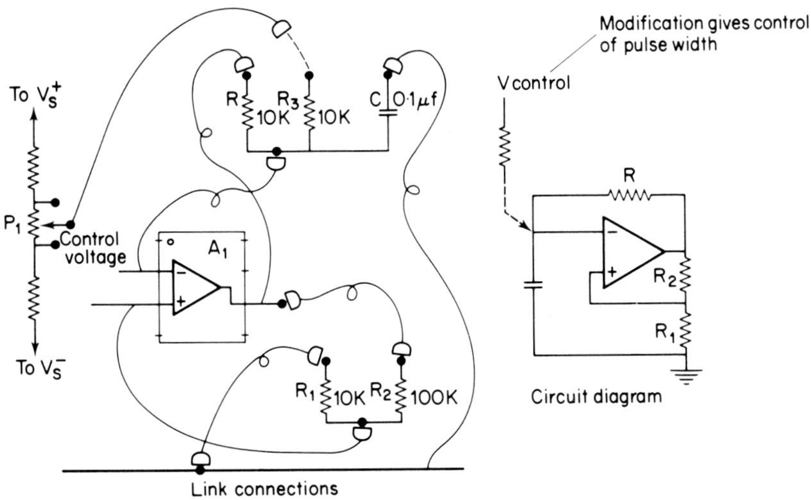

Figure 45 A free-running multivibrator

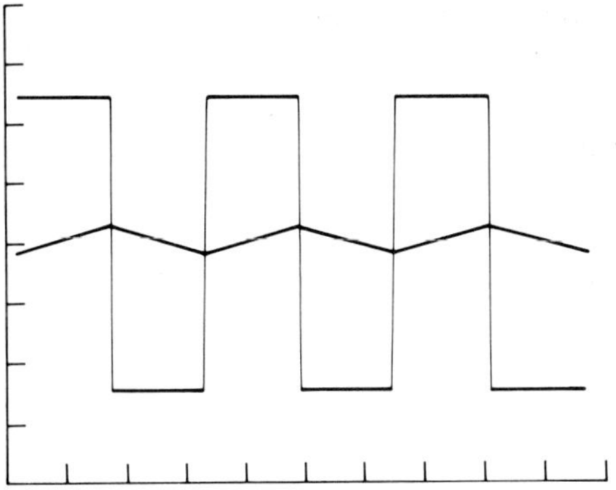

Figure 46 Multivibrator waveforms (vertical sensitivity, 5 V/division; horizontal sensitivity, 0.1 ms/division)

waveforms with a dual-trace oscilloscope. Typical waveforms at the amplifier output and inverting input are shown in Figure 46. Relate the measured waveform amplitude and time periods to the discussion of circuit action given in the companion text.

Control of pulse width

Add the modification to the circuit which is indicated by the broken lines shown in Figure 45. Note the effect on circuit waveform of varying the control voltage provided by potentiometer P_1. Explain the observed effects.

Exercises 13

1 Observe the effect on the circuit waveforms of changing the amplifier power supply voltages. Change the positive supply from +15 V to +10 V. Reset the positive supply to +15 V and change the negative supply from −15 V to −10 V. Explain the observed effects (see equations 7.3 and 7.4 of the companion text).

2 Observe the effect on the circuit waveforms of changing the resistor ratio R_2/R_1. use $R_1 = R_2$ of say 10 kΩ. Explain the observed effects.

3 Assuming that the operational amplifier used in Figure 45 has output saturation limits ± 10 V, find component values which will cause the sawtooth waveform to have a peak-to-peak size of 4 V and a frequency of 1 kHz (use $C = 0.01\ \mu$F).

Experiment 14
A triangular/square waveform generator

In this experiment the student is asked to examine the behaviour of a waveform generator circuit. The circuit is formed by combining an operational integrator with a regenerative comparator. Waveform generator circuits are discussed in Section 7.4 of the companion text.

Experimental procedure

Make the link connections shown in Figure 47a. Measure the waveforms which appear at the output of each amplifier. Typical waveforms obtained in the experiment are shown in Figure 47b. Relate your observations to the discussion of circuit action which is given in the companion text. In particular you should make a quantitative comparison between experimentally measured waveform amplitudes, slopes and time periods and the waveform shown in Figure 7.19 of the companion text.

Use the potentiometer on the experimental board as a means of applying offsets to the operational amplifiers (as shown in Figure 7.20 of the companion text). Observe the effect on circuit waveform of these offsets and relate your observations to the discussion given in Section 7.4.2 of the companion text.

Exercises 14

1 Observe the effect on circuit waveforms of

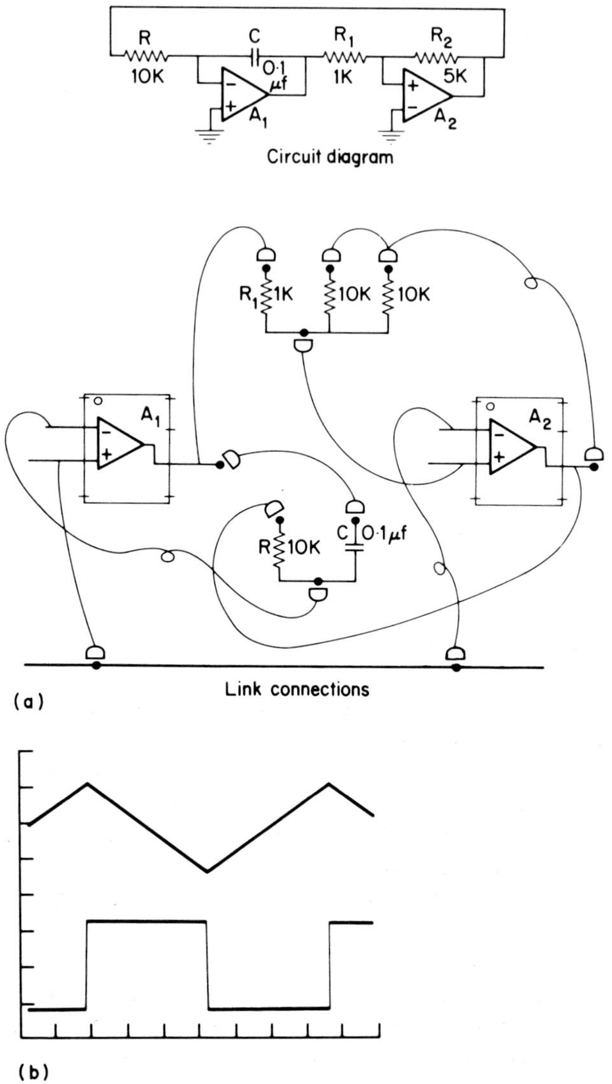

Figure 47 An integrator and a regenerative comparator form a triangular/square wave generator: (a) circuit diagram and link connections; (b) typical waveforms (vertical sensitivity, upper trace 2 V/division, lower trace 10 V/division; horizontal sensitivity, 0.1 ms/division)

(a) changing the value of R to 100 kΩ;
(b) making $R_2 = 100$ kΩ, $R_1 = 10$ kΩ.
Explain the observed effects in terms of the principles underlying the circuit action.

2 Assuming that the operational amplifiers used in Figure 47 have output saturation limits ± 10 V, find component values which will make the circuit produce a triangular wave of peak-to-peak value 10 V and frequency 1 kHz (use $C = 0.01$ μF).

Experiment 15
A simple voltage-to-frequency converter

In this experiment the student is asked to examine the action of a simple voltage-to-frequency converter circuit. The circuit is made by combining an operational integrator with a regenerative comparator; it is described in Section 8.13 of the companion text.

Experimental procedure

Make the link connections shown in Figure 48. Potentiometer P_1 is used as a means of applying a variable d.c. input signal; the voltage applied should be measured with a d.c. voltmeter. Observe the waveform which appears at the output of each amplifier. Typical waveforms obtained in the experiment are shown in Figure 48b.

Measure the frequency of oscillations for various values of the d.c. input signal and plot a graph of frequency against input voltage. The frequency may be obtained as the inverse of the waveform period as measured by the oscilloscope. Alternatively the frequency may be measured directly with a counter/timer frequency meter, if such an instrument is available.

Observations made in the experiment should be related to the discussion of circuit action which is given in the companion text.

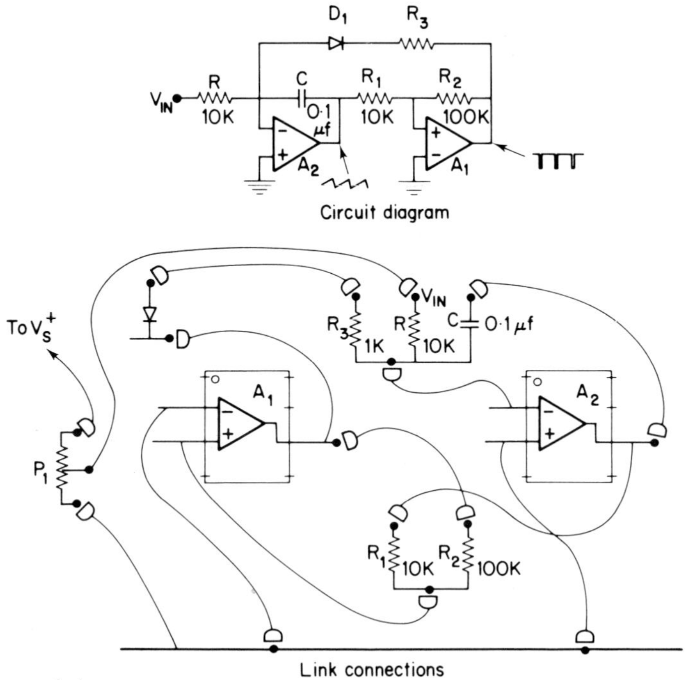

(a)

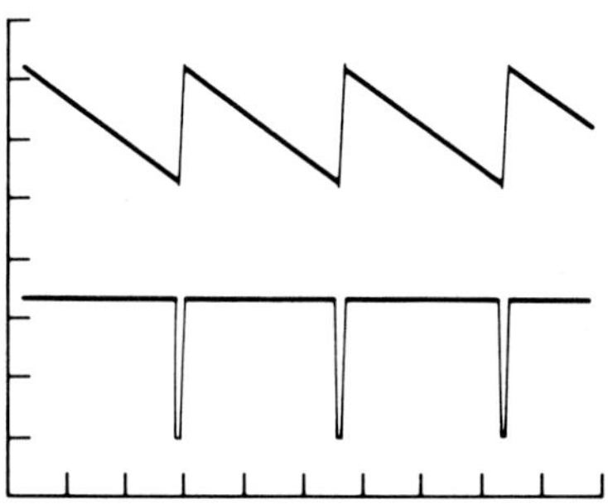

(b)

Figure 48 An integrator and a regenerative comparator form a simple voltage-to-frequency converter: (a) circuit diagram and link connections; (b) typical waveforms (vertical sensitivity, upper trace (A_2 output) 2 V/division, lower trace (A_1 output) 10 V/division; horizontal sensitivity, 50 μs/division)

Experiment 16
Pulse width modulation

This experiment suggests a procedure for investigating the action of a simple pulse width modulator system. The system is described in Section 8.15.2 of the companion text.

Modulating action is obtained by using an operational amplifier as a comparator. A sinusoidal test signal is applied as a modulating signal to the non-inverting input terminal of the comparator. A triangular wave, acting as a carrier signal, is applied to the inverting input terminal. The comparator output switches between its positive and negative saturated output values every time the signals at its input terminals pass through equality. As a result of this action the comparator produces a square-wave output signal; the width of the square wave is modulated in accordance with the amplitude of the modulating sine wave.

Experimental procedure

The experimental procedure adopted depends upon the test waveform generators which the experimenter has available. Assuming that

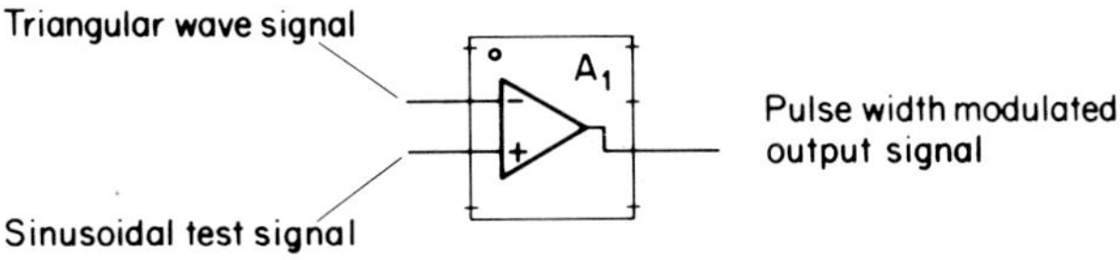

Figure 49 An operator amplifier used as a pulse width modulator

a triangular waveform generator and a sinusoidal signal generator are available, then signals are simply applied to the amplifier input terminals as indicated in Figure 49. The amplitude of the sinusoidal signal is adjusted so that it is slightly smaller than the amplitude of the triangular wave. Input and output signals are monitored with an oscilloscope. Typical waveforms obtained in the test are given in Figure 50.

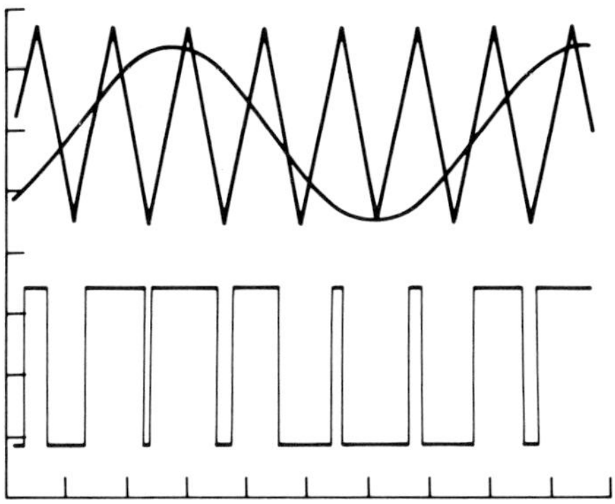

Figure 50 Waveforms obtained in the pulse width modulator experiment: upper trace, modulating signal and triangular carrier signal (vertical sensitivity, 2 V/division; horizontal sensitivity, 0.5 ms/division); lower trace, pulse-width-modulated signal (vertical sensitivity, 10 V/division; horizontal sensitivity, 0.5 ms/division)

Note. It is necessary to transfer the oscilloscope test leads between circuit points in order to obtain the three waveforms shown. Since the sinusoidal modulating signal and triangular carrier signal are not in synchronism the lower trace does not correctly indicate the instantaneous time relationships between the signals.

If a triangular waveform generator is not available to the experimenter he may produce a triangular wave by integrating a square-wave test signal. The a.c. integrator circuit described in Experiment 9 and given in Figure 27 may be used for this purpose. Alternatively the triangular/square waveform generator circuit examined in Experiment 14 may be assembled and used to provide a triangular test signal.

Experiment 17
Sinusoidal oscillators

Operational amplifier circuits which generate sinusoidal oscillations are examined in this experiment. A Wien bridge oscillator and a quadrature oscillator circuit are investigated. A discussion about sinusoidal oscillator circuits is given in Section 7.3 of the companion text.

Wien bridge oscillator

Make the link connections shown in Figure 51. Positive feedback

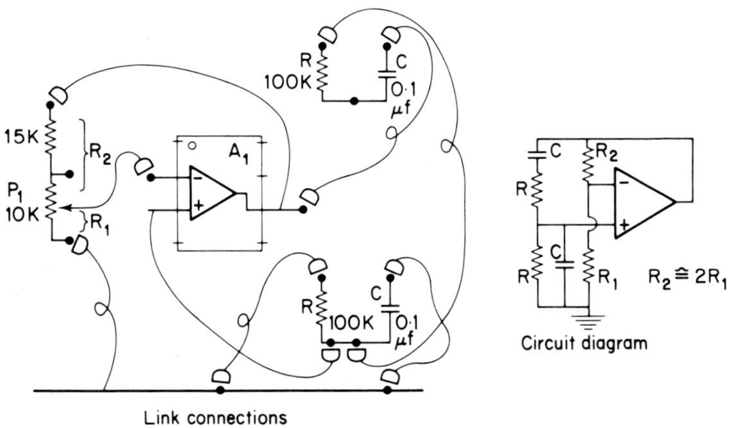

Figure 51 A Wien bridge oscillator

is applied to the operational amplifier via the resistor–capacitor path between its output and non-inverting input terminal. Negative feedback is applied via the path R_2, R_1 between its output and inverting input terminal. The relative values of resistors R_2 and R_1 are determined by adjustment of potentiometer P_1.

Connect the oscilloscope test leads to the output of the amplifier in order to observe any signal which appears there. Adjust potentiometer P_1 in order to obtain a sinusoidal waveform which is just maintained. Use a resistance meter to measure the ratio R_2/R_1 which is established by this adjustment. Measure the frequency of oscillations.

Verify that the frequency of oscillations is determined by the relationship $f = 1/2\pi CR$ and that oscillations are maintained when R_2/R_1 is approximately 2. Notice that if R_2 is made appreciably greater than $2R_1$ a square-wave oscillation is produced and if R_2 is made less than $2R_1$ oscillations decay and cease.

A quadrature oscillator

A quadrature oscillator produces sine and cosine waves; that is, it

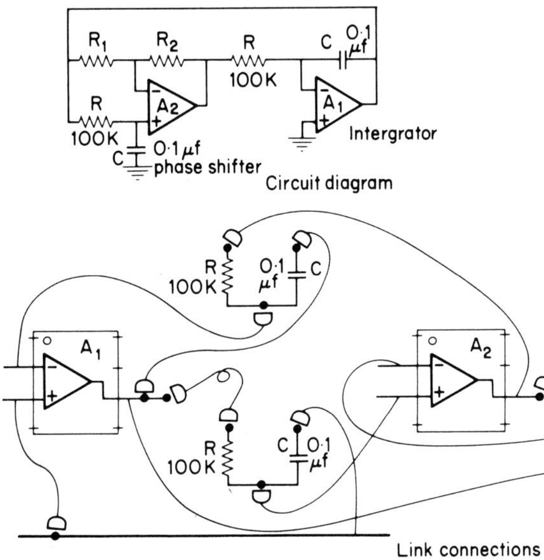

Figure 52 A quadrature oscillator formed by an integrator and a phase shifter

produces two sinusoidal signals with a 90° phase difference between them. Figure 52 shows the link connections required to arrange a form of quadrature oscillator circuit on the experimental board. The circuit uses two operational amplifiers, one connected as an integrator and the other as a phase shifter.

Connect the circuit and use a dual-trace oscilloscope to observe the signals appearing at the output terminal of each amplifier. Adjust potentiometer P_2 until an oscillation with a sinusoidal waveform is just maintained. Use a resistance meter to measure the ratio R_2/R_1 which is established by this adjustment of P_2. Measure the frequency of oscillations.

Verify that the frequency of oscillations is determined by the relationship $f = 1/2\pi CR$. Oscillations grow in amplitude to give a square-wave oscillation when $R_2 > R_1$ and decay when $R_2 < R_1$.

Observations made with both circuits should be related to the discussion given in the companion text.

Experiment 18
Precise rectifier circuits

Near ideal diode behaviour (precise diode action) can be obtained by using an operational amplifier to reduce the effects of the non-linear forward conduction characteristics which are exhibited by real diodes. Several different precise diode circuit configurations are examined in this experiment. The circuits given are discussed in Section 8.10 of the companion text.

Precise half-wave and full-wave rectifier circuits

The link connections on the experimental board required to make precise half-wave and full-wave rectifier circuits are shown in Figure 53.

Make the link connections for the half-wave circuit first (the connections associated with amplifier A_1). Apply a sinusoidal input test signal of frequency say 400 Hz and with amplitude a volt or so. Observe the waveforms at the output of A_1 and at the anode of diode D_1 (point X). A half-wave rectified output appears at X. Typical waveforms are shown in Figure 54. Trace a is the input wave, trace b the waveform at the output of A_1 and trace c the waveform at X.

Add the additional link connections required to change the half-wave circuit into a full-wave circuit. These are the connections associated with amplifier A_2. A_2 is arranged as an inverting adder. Again apply a sinusoidal input test signal and adjust potentiometer

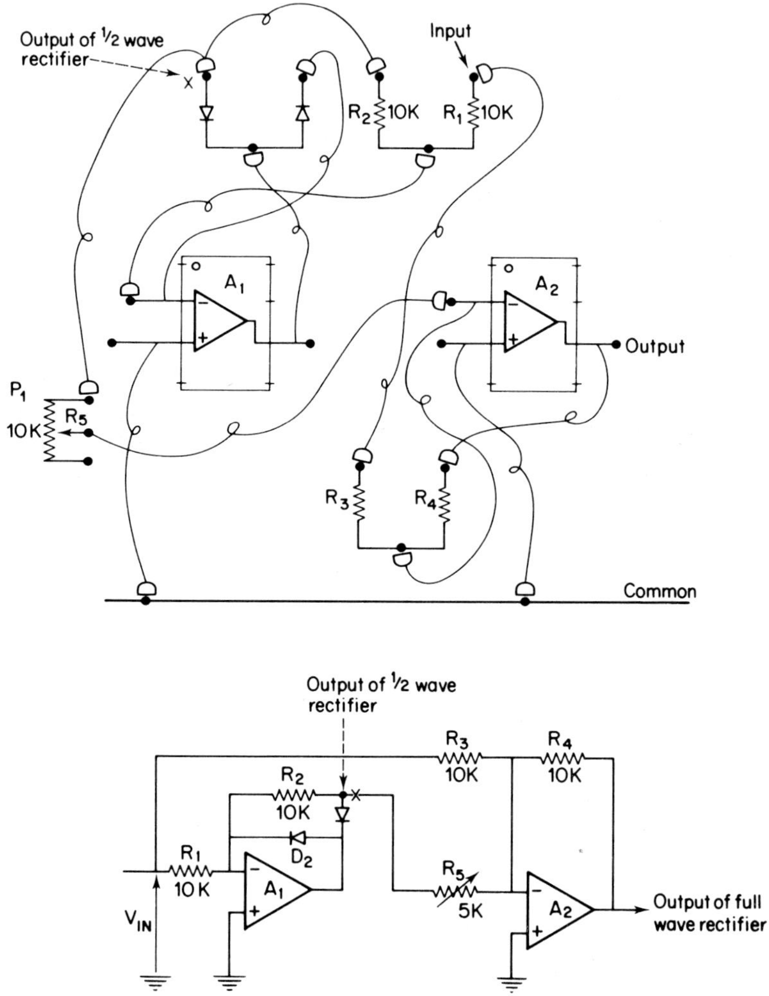

Figure 53 Ideal half-wave and full-wave rectification

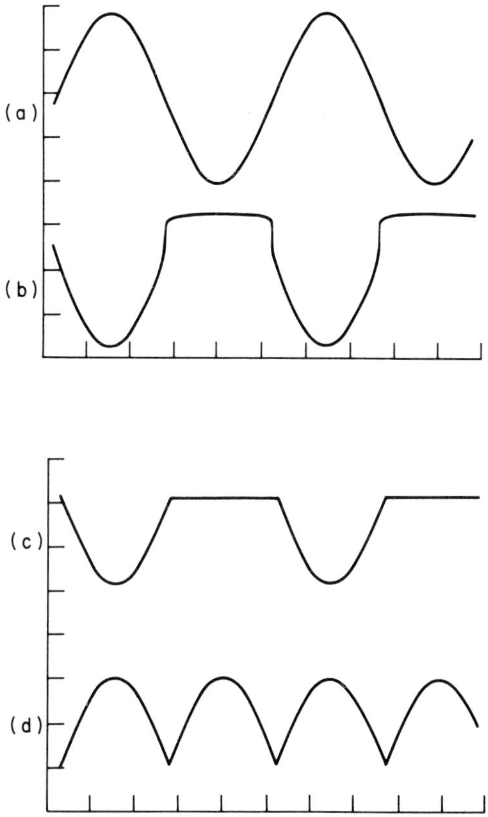

Figure 54 Waveforms obtained with precise half-wave and full-wave rectifier circuits: (a) input signal; (b) A_1 output; (c) output at X; (d) A_2 output

P_1 in order to obtain a full-wave rectified signal at the output of A_2. A trace of the type shown in Figure 54d should be obtained.

Investigate the range of signal frequencies and amplitudes for which satisfactory operation of the circuit is obtained. You should attempt to interpret the performance limits of the circuit in terms of the performance parameters of the operational amplifiers.

Precise full-wave rectifier circuits are often referred to as 'absolute-value circuits'. An absolute-value circuit produces a unipolar output

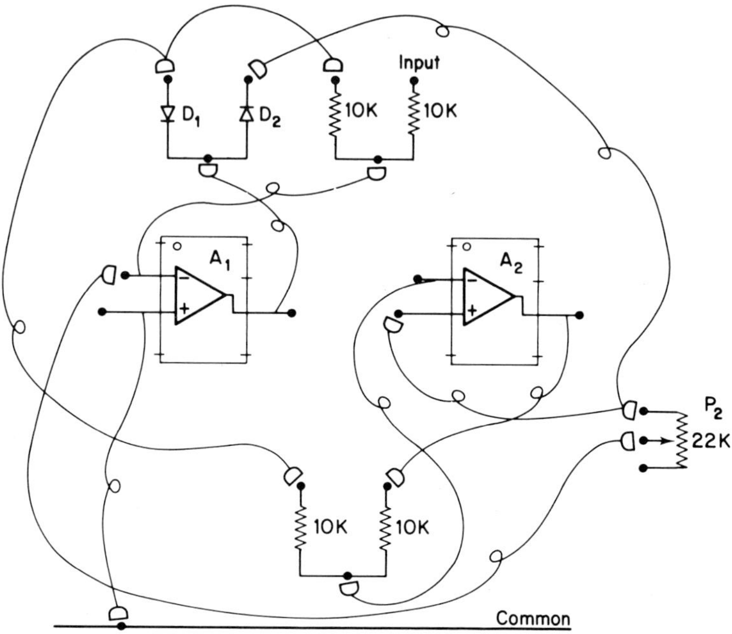

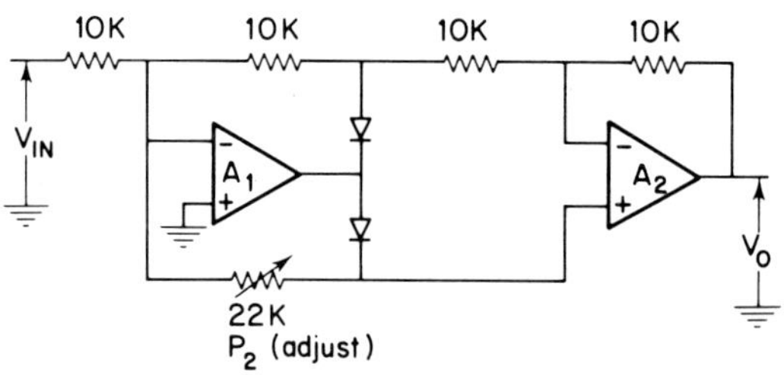

Figure 55 An absolute-value circuit requiring five equal-value resistors (adjust P_2 for full-wave rectified output at A_2)

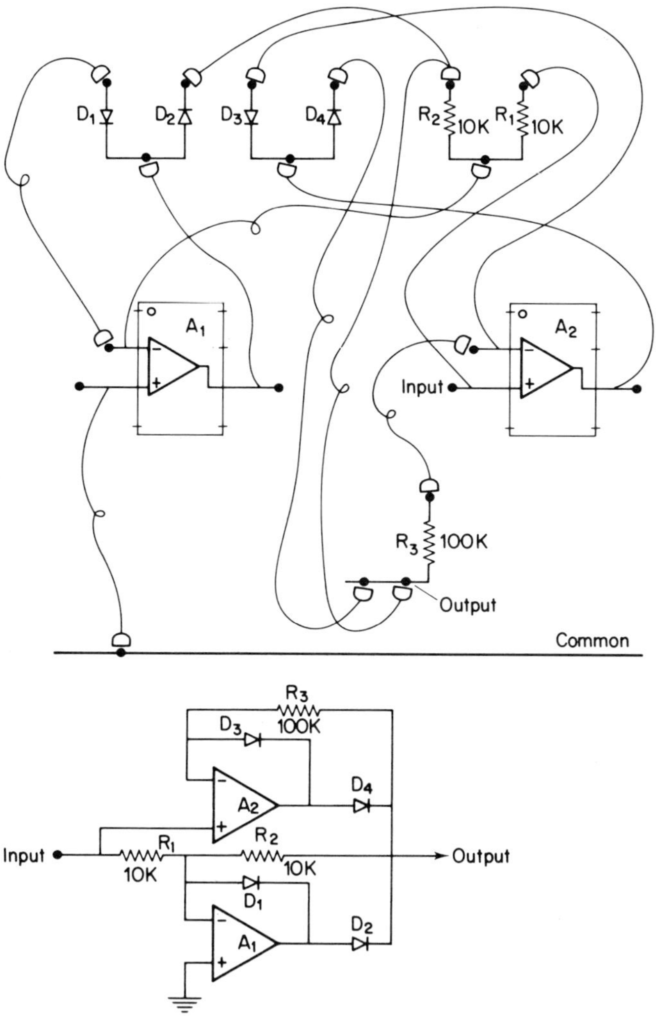

Figure 56 An absolute-value circuit requiring only two matched resistors

signal with magnitude equal to the magnitude of the input signal which is applied to it. The polarity of the output remains unchanged as the input polarity changes. Other examples of absolute-value circuits which can be linked up on the experimental board are given in Figures 55 and 56.

Precise level selection

Link connections of the experimental board required to form a precise two-input upper-level selector are shown in Figure 57. An

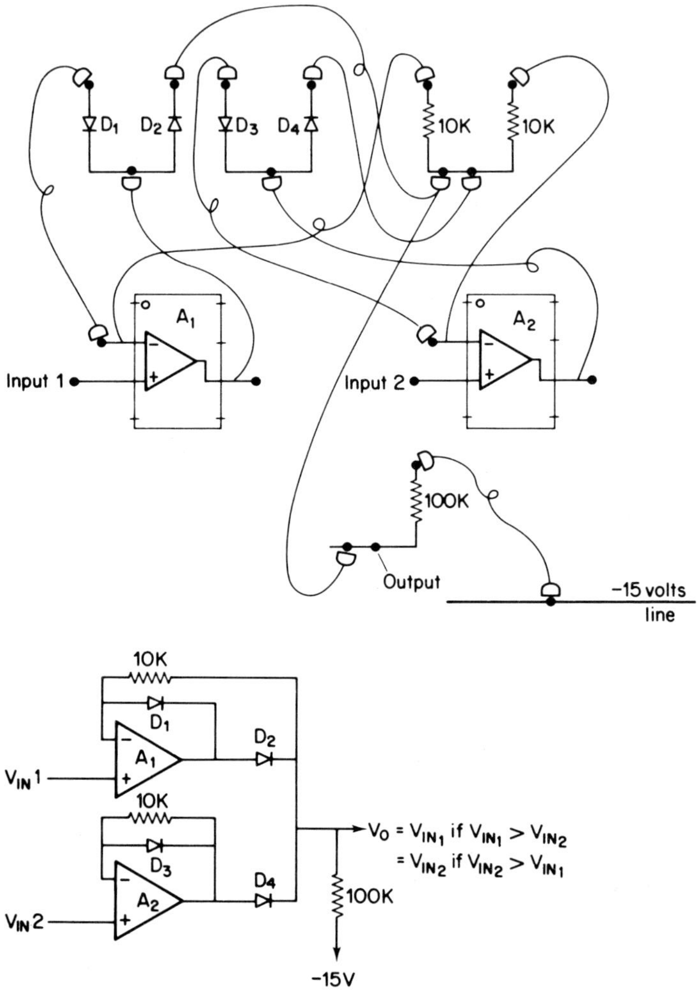

Figure 57 A precise upper-level selector

upper-level selector provides an output signal which is equal to the most positive of the several input signals which are applied to it. The system in Figure 57 has only two inputs. It may be extended to accommodate any number of inputs by adding an extra operational amplifier and associated diodes for each extra input.

The action of the system can be investigated by applying two input test signals and observing the output with an oscilloscope. Apply say a d.c. signal to input 2 (derived from potentiometer P_2 connected across the supply rails) and a sinusoidal signal to input 1. Vary the signal applied to input 2.

Consider the modification to the system required to make it function as a lower-level selector. Make the modifications and investigate the action of the system as before.

Precise peak detection

A peak detector measures the maximum positive value which a varying signal has during a prescribed time interval (since the last reset). The link connections of the experimental board which are required to form a precise low-drift peak detector are shown in Figure 58.

Apply an input signal to the circuit, say a signal derived from potentiometer P_2 connected across the supply rails. Notice that as the input signal is increased the output follows it. However, when the input signal is decreased the output 'holds' at the maximum value of the input. The output is reset by shorting the hold capacitor C.

In the peak hold mode the output slowly decays (drifts). Drift in the peak hold mode is mainly due to the bias current drawn by amplifier A_2:

Drift rate $\approx I_b/C$

FET input operational amplifiers have a very small bias current and give the circuit a low drift rate.

Exercises 18

1 What changes are required in the arrangement shown in Figure 58 in order to make it function as a valley detector (a circuit which

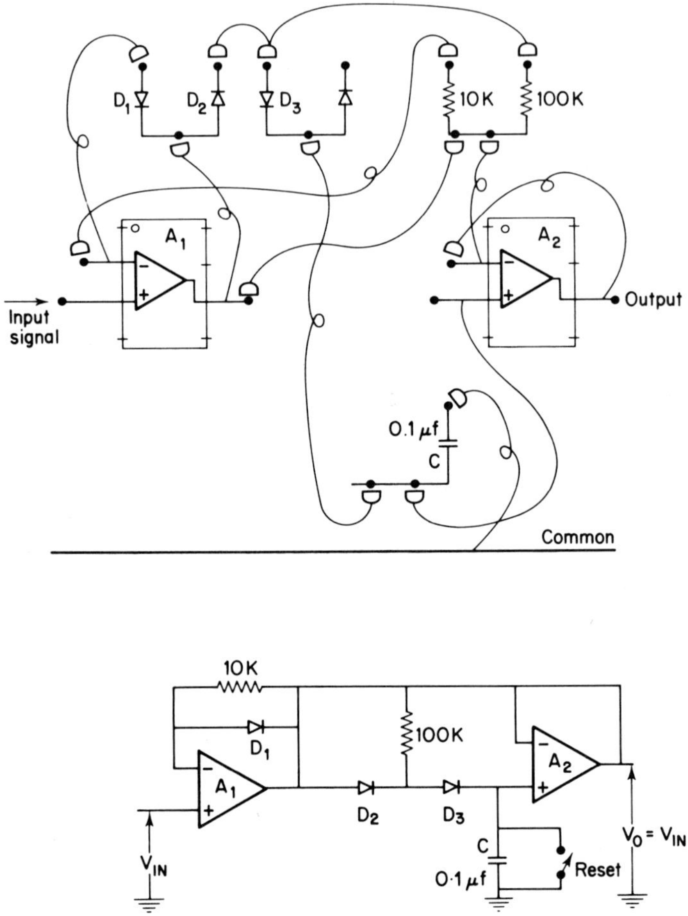

Figure 58 A precise low-drift peak detector

detects the most negative value of an input signal)? Make these changes and verify the action of the circuit.

2 If the operational amplifier A_2 in Figure 58 has a bias current of 100 pA, what output drift rate do you expect to observe in the peak hold mode?

Experiment 19
A switched gain polarity circuit

An FET can be used to switch the closed-loop signal gain of an operational amplifier between the values plus and minus unity. The action of a switched gain polarity circuit is examined in this experiment; the circuit is discussed in Section 8.12 of the companion text, and the link connections which are required to assemble the circuit are given in Figure 59.

The FET, included on the experimental board, is connected between the non-inverting input terminal of amplifier A_1 and the common line. The FET exhibits a low resistance between its drain and source terminals when it has a zero bias voltage applied between its gate and source terminals. Under these conditions the non-inverting input terminal of the amplifier is effectively shorted to the common line and the circuit acts as an inverter with gain minus unity.

A negative bias voltage applied between gate and source terminals, if it is of sufficient size, cuts off the FET and makes it act like an open switch between its drain and source terminals. There is then the same input signal applied to both the inverting and non-inverting input terminals of the operational amplifier and the circuit acts as a unity-gain follower (gain $+1$).

Experimental procedure

Make the link connections shown in Figure 59. The FET is switched off by making a link connection between the point X and the power

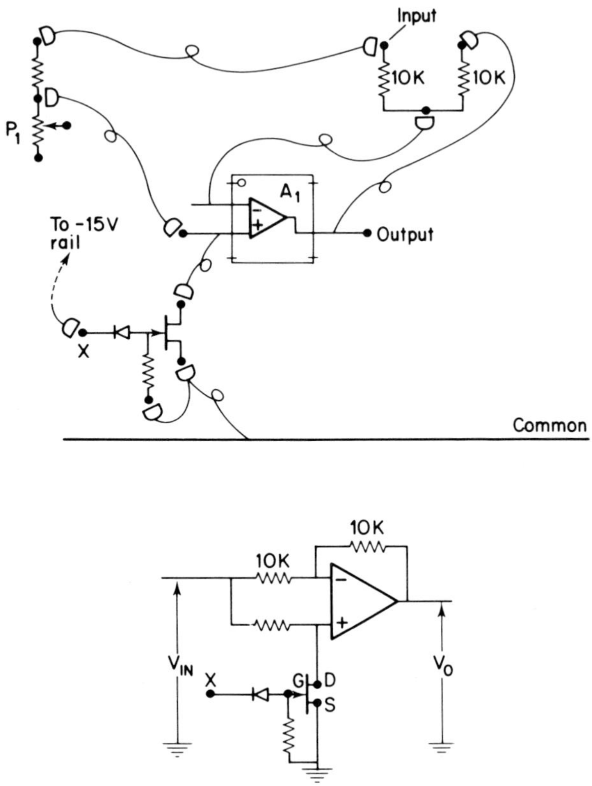

Figure 59 A circuit with its gain switched between +1 and −1

supply negative rail; it is switched on by removing this link connection.

Apply a low-frequency sinusoidal signal (amplitude say 5 V) as an input test signal and monitor the input and output signals with a dual-trace oscilloscope. Input and output waveforms will be in phase (gain +1) when the FET is switched off, and they will be in antiphase (gain −1) when the FET is switched on.

Experiment 20
A phase-shifting circuit

Occasions arise when it is necessary to adjust the phase of a sinusoidal signal without changing its amplitude. An operational

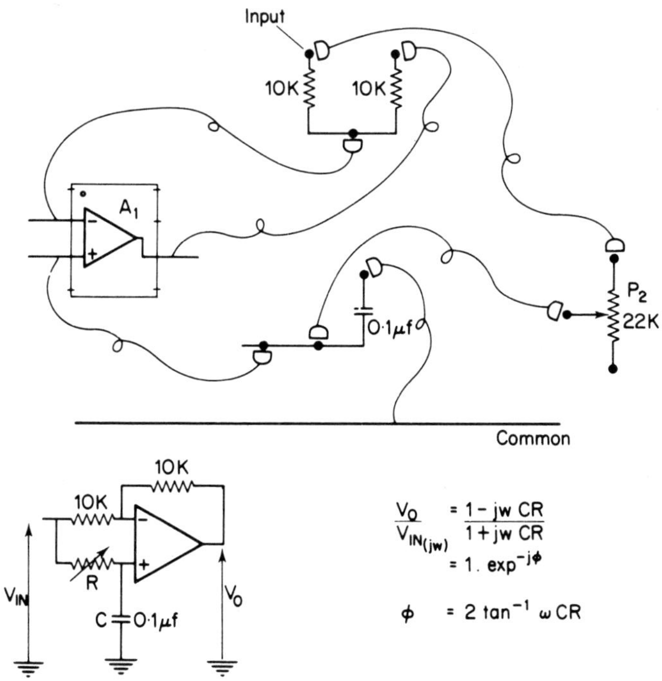

$$\frac{V_O}{V_{IN(jw)}} = \frac{1 - jw\,CR}{1 + jw\,CR} = 1.\ \exp^{-j\phi}$$

$$\phi = 2\tan^{-1}\omega CR$$

Figure 60 A phase-shifting circuit

amplifier circuit for this purpose is discussed in Section 8.7 of the companion text; a practical form of the circuit which can be assembled on the experimental board is shown in Figure 60.

Experimental procedure

Make the link connections shown in Figure 60. Apply a sinusoidal input test signal (amplitude say 5 V, frequency say 1 kHz) and monitor the input and output signals with a dual-trace oscilloscope. Vary the setting of potentiometer P_2 from zero to its maximum resistance (22 kΩ) and note the effect that this has on the phase relationship between the input and output signals.

The phase shift produced by the circuit is determined by the relationship

$$\phi = -2\tan^{-1} 2\pi fCR$$

There is zero phase shift when the potentiometer is set at its zero setting and a maximum phase shift (phase lag) of

$$2\tan^{-1}(2\pi 10^3 \times 0.1 \times 10^{-6} \times 22 \times 10^3) \approx 172^\circ$$

when the potentiometer is set to its maximum value 22 kΩ. Phase shift can be estimated from oscilloscope traces as shown in Figure 61.

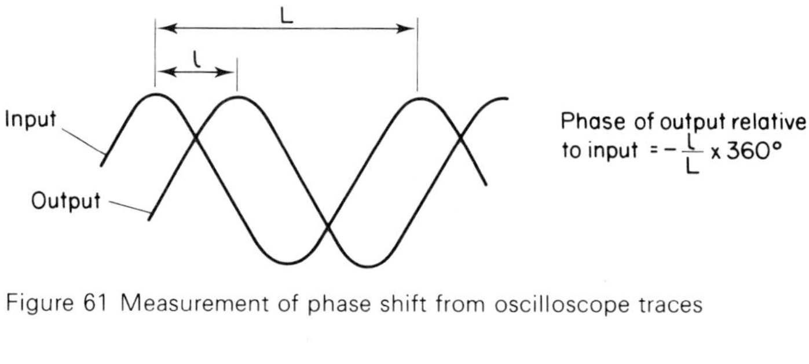

Figure 61 Measurement of phase shift from oscilloscope traces

Experiment 21
A phase-sensitive detector system

Introduction

Phase-sensitive detection is a synchronous rectifying process; a phase-sensitive detector (PSD) gives a unidirectional output signal (a rectified output signal) for alternating input signals which are in synchronism with an alternating reference signal which is also applied to the detector. All input signals that are not synchronous with the reference signal give rise to alternating signals at the detector output.

The functional operation which is performed by a PSD may be described in terms of a multiplication process. The output signal produced by a PSD is determined by the product of the input signal and the reference signal. Most PSDs use a square-wave reference signal. The action of the PSD is to cause the square wave to multiply the input signal alternately by +1 and −1; waveforms illustrating the process are shown in Figure 62. A sinusoidal input signal is shown in Figure 62; when the input signal and reference signal are at the same frequency and are in phase (or 180° out of phase) full-wave rectification occurs. The average value $\bar{V}_o$ of the output has a maximum positive value when the input signal and the reference signal are in phase, a maximum negative value when input and reference are in antiphase and is zero when input and reference are 90° out of phase. The output signal is an alternating signal, which averages to zero, when the input signal and reference signal are unrelated. Averaging is performed by a low-pass filter connected to the output of the PSD.

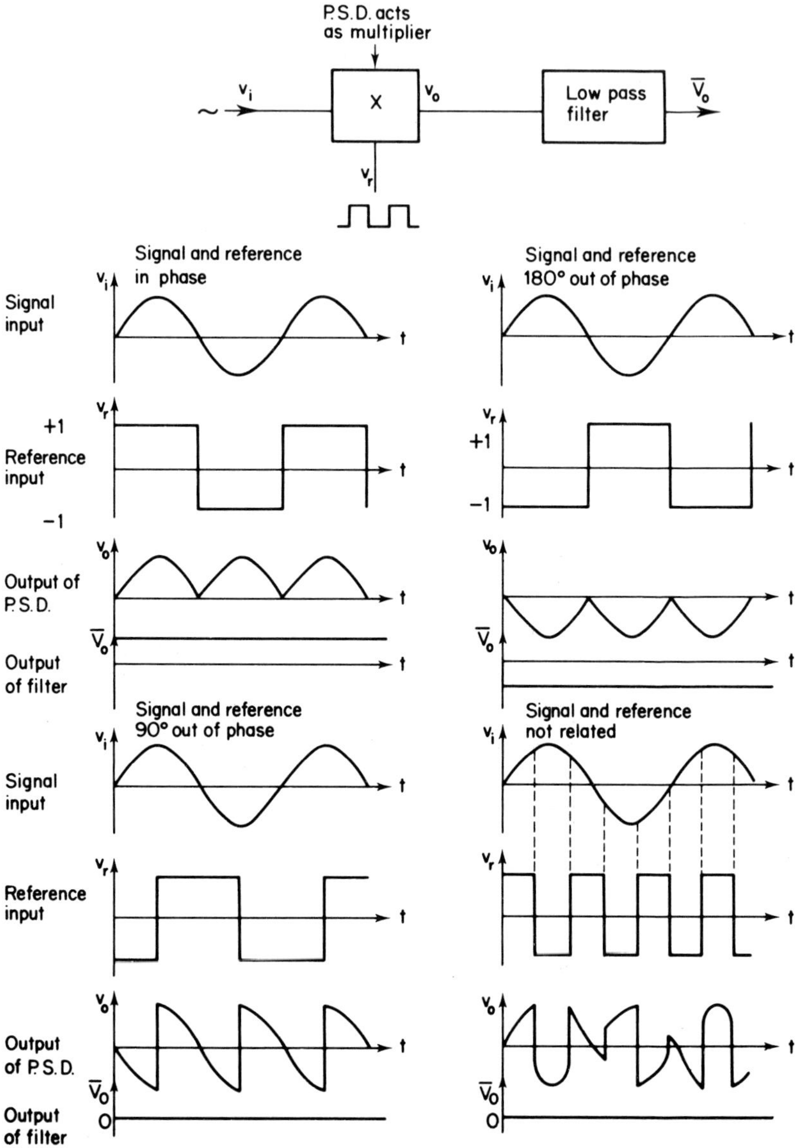

Figure 62 Functional operation of a PSD

An experimental PSD system

The switched gain polarity circuit and the phase-shifting circuit (given in the two previous experiments) can be combined with a squaring circuit and a low-pass filter to provide a PSD system suitable for experimental evaluation. This system is assembled on two of the experimental boards and is shown in block schematic form in Figure 63. Details of the link connections which are required on each board are given in Figures 64 and 65.

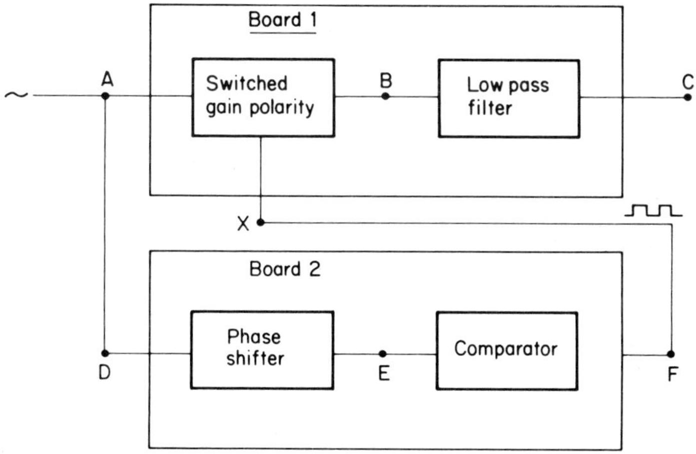

Figure 63 A block schematic for an experimental PSD system

Experimental procedure

Make the link connections shown in Figures 64 and 65 on two separate experimental boards. Interconnect the two boards as shown in Figure 63: point F (the output of the comparator operational amplifier) to point X (the gate of the switch FET); point A (the input to the PSD) to point D (the input to the phase-shifting circuit).

Apply a sinusoidal signal (amplitude say 5 V, frequency say 1 kHz) as an input test signal and monitor the signals at points A and B with a dual-trace oscilloscope. Observe the effect of varying potentiometer P_2 on board 2 (varying the phase angle between input signal

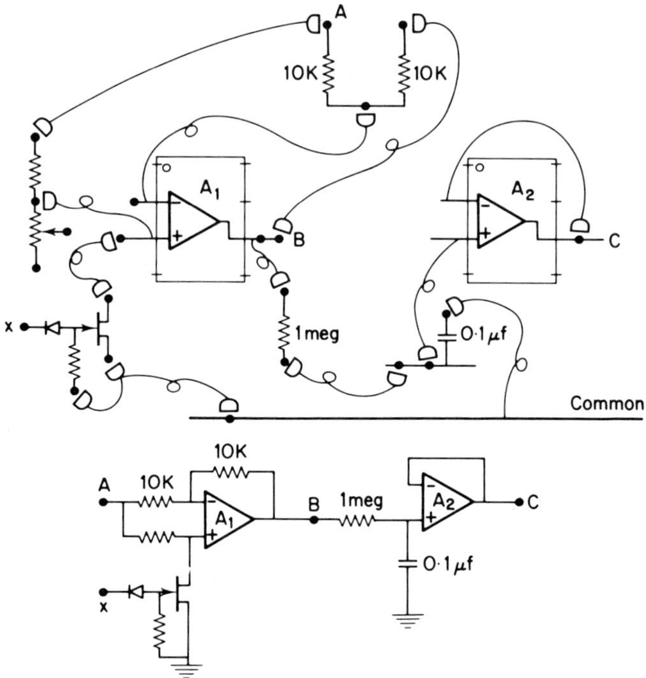

Figure 64 Board 1 of PSD system: switched gain circuit and low-pass filter

and reference); start with P_2 set at zero and increase it to its maximum value. Relate your observations to the functional operation of the system, as described in the introduction and as illustrated by the waveforms given in Figure 62.

Use a d.c. voltmeter to measure the output signal produced by the low-pass filter. Record values of this d.c. output for various settings of the phase shift. Verify the equation $\bar{V}_o = (2/\pi)V_{in}\cos\phi$, where V_{in} is the amplitude of the sinusoidal input signal and ϕ is the phase shift introduced by the phase-shifting circuit. Phase shift may be estimated as in the previous experiment.

The sinusoidal test signal used in the above experimentation acts both as an input signal to the PSD and as the reference signal. It should be noted that the multiplication by $+1$ and -1 which is per-

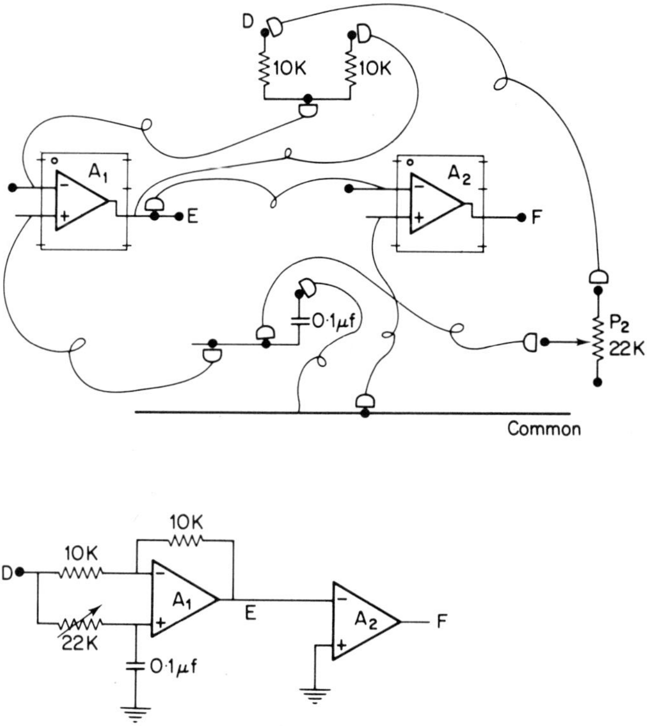

Figure 65 Board 2 of PSD system: phase-shifting circuit and zero reference level comparator (squarer)

formed by the PSD is in response to the lower and upper levels respectively of a square wave. This square wave is derived from the phase-shifted reference signal by the action of the zero reference level comparator (amplifier A_2, board 2, acting open loop); it is applied to the gate terminal of the FET switch (board 1).

PSDs are used as a means of recovering signals which are obscured by noise. All input signals not synchronous with the reference signal (this includes all random noise signals) cause the PSD to give an alternating output signal which the low-pass filter averages to zero (provided that its cut-off frequency is low enough).

The noise-rejection property of a PSD can be demonstrated by using two separate input test signals, one acting as a reference signal and applied to the system at point D, the other acting as the input signal to the PSD and applied at point A. Observe the waveforms at points A and B and the reading of the d.c. voltmeter at C whilst varying the frequency of the signal applied to point A. The frequency of the reference signal applied at point D should be held constant (say at around 500 Hz).

A PSD gives a non-zero average output signal for a signal at the reference frequency and also for input signals with frequencies equal to odd harmonics of the reference signal frequency:

$$\bar{V}_o = (2/3\pi)V_{in}\cos\phi$$

for a sinusoidal input signal with $f_{in} = 3f_r$;

$$\bar{V}_o = (2/5\pi)V_{in}\cos\phi$$

for an input signal with $f_{in} = 5f_r$. Some care in frequency adjustment is required in order to observe the response of the system to odd harmonics of the reference signal. The two separate input test signals are in no way synchronized to each other and small frequency drifts tend to destroy any exact setting of the ratio f_{in}/f_r.

Appendix

Using a personal computer to learn about operational amplifiers

Learning through computers is by no means a new development. The techniques of Computer Assisted Learning, C.A.L. and Computer Assisted Instruction, C.A.I. have been around since the late 1960s. Until recently the cost of the necessary computer hardware has been such as to prevent the widespread use of these techniques. The comparatively low cost of today's microcomputers has removed economic considerations as the main barrier preventing a wider dissemination of C.A.L. techniques.

Science uses conceptual models to explain the behaviour of 'real world' things and events. Quantitative behaviour is expressed in the form of mathematical equations (mathematical models) and extensive use is made of diagrams and graphs in order to gain insight into the nature of a theoretical model. A unique advantage of the computer as an educational tool lies in the capacity to embody a theoretical model into a computer program. A well designed program allows the user to test out the model by altering to his own specifications the various conditions and parameters of the model. Insight and learning are enhanced by the user being able to get immediate answers and feedback as to the effect of the changes he has introduced on the model's behaviour.

The high resolution graphics capabilities of modern micros allows a well-defined screen presentation of the diagrams and graphs which are used to represent the model's behaviour. A micro and TV display can be thought of as an alternative to an overhead or slide projector, but together they have the enormous advantage of allowing the user to interact directly with the screen presentation and influence what is displayed in a manner determined by the particular model incorporated into the computer program.

A series of modelling type C.A.L. programs have been written to support this manual and its companion text. Two examples of the screen presentations generated by the programs are illustrated by the printouts reproduced in Figures A1 and A2.

Figure A1 is produced by a program which deals with the basic feedback principles underlying the action of operational amplifier inverter and follower circuits; circuits which are experimentally

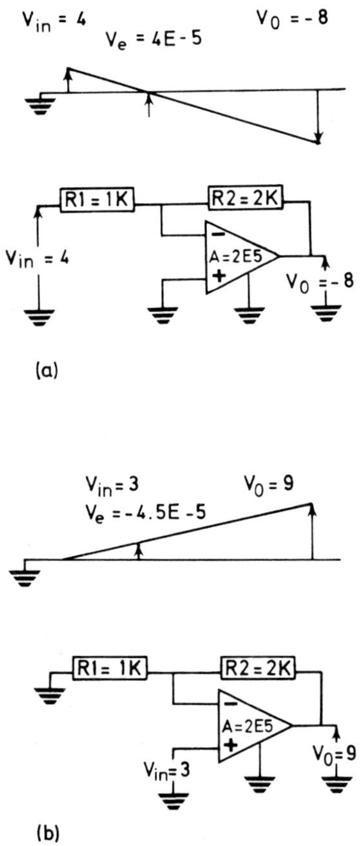

Figure A1 Computer printouts giving lever analogy for operation of inverter and follower

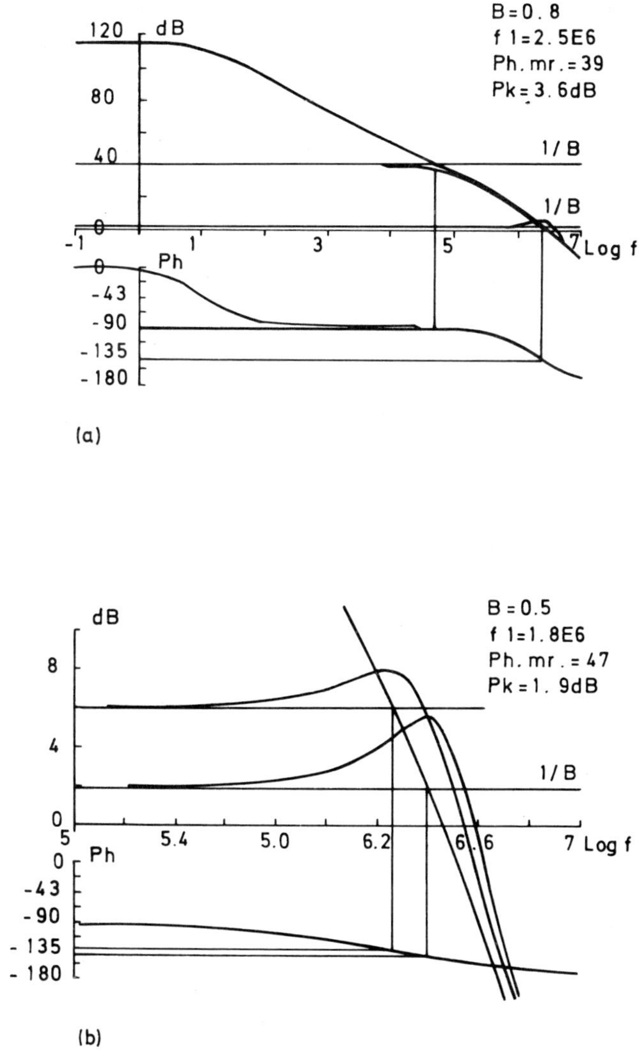

Figure A2 Computer printouts of Bode plots

investigated in Experiment 1 of this manual. A lever analogy for the operation of the circuits is presented in the program.

Parameters of the model which can be generated by the program or entered by the program user are, resistor values, the amplifier open loop gain, the amplifier power supply values and the size and polarity of the applied input signal. The program provides insight into the importance of the feedback fraction and the loop gain in determining closed-loop behaviour. The effect of power supplies in determining amplifier output saturation limits is incorporated in the model.

The graphs shown in Figure A2 are examples of screen presentations generated by a program dealing with the frequency response characteristics of amplifiers; the subject matter is dealt with in Experiments 3, 4, 5 and 6 of this manual. The program allows for an investigation of the effects of a circuits feedback fraction on closed-loop bandwidth, phase margin and gain peaking. Figure A1(b) illustrates a scaling capability which allows a magnified presentation of a small section of a graph.

Programs that are available are written for use with the B.B.C. microcomputer; they will be made available for use with other computers. Full details of the subject matter covered by the programs may be obtained from:

S.C.I.C.A.L. Software,
13 Kingsmead Road North,
Oxton,
Birkenhead.
Merseyside L43 6TB.

Answers to exercises

Exercises 1

1 (a) Circuit of Figure 4 with $R_1 = 1$ kΩ,$R_2 = 100$ kΩ;
1 (b) Circuit of Figure 3 with $R_1 = 1$ kΩ and $R_2 = 22$ $k\Omega$ variable resistor;
(c) Circuit of Figure 5 with $R_f = 10$ kΩ, $R_1 = 1$ kΩ and $R_2 =$ 100 kΩ.

Exercises 2

1 $V_{io} = 2.5$ mV; $I_b^+ = 60$ nA; $I_b^- = 70$ nA.
2 (a) 27.9 mV; (b) 30 mV.

Exercises 3

2 (a) 10 (20 dB); 25 (28 dB); 250 (48 dB);
(b) 12.6 Hz.

Exercises 4

1 (a) 1.5×10^6 Hz, 0.22 μs;
(b) 3×10^5 Hz, 1.1 μs;
(c) 7.5×10^5 Hz, 0.44 μs.
2 25°.

Exercises 5

1 (a) 7.96 kHz; (b) 15.9 kHz; (c) 31.6 kHz.
2 9.42.

Exercises 6

1 600 pF; 0.33 V/μs.
2 13.3 pF; 800 kHz; 1.2 V/μs.

Exercises 7

1 Circuit of Figure 19 with $R_1 = R_2$ (say 10 kΩ).
2 Circuit of Figure 20 with $R_1 = R_2$ (say 10 kΩ).

Exercises 8

1 +4 V; +2 V; 0 V; −1.4 V; −2 V; $V = \sqrt{10} \times 10^{-1}$; $K = 5.33$.
2 $I_{o_1}/I_{o_2} = 1.2/1$.
3 $K \cong 60 \times 10^{-3} \dfrac{R_3 + R_1}{R_3}$; $I_{ref} = \dfrac{V_s}{R_2}$; $K = 5$; $I_{ref} = 10^{-6}$ A; $R_4 =$ 82.3 kΩ; $R_2 = 15$ MΩ.

Exercises 9

1 (a) 3 mV/s; (b) 15 V/s.
2 ±5 V peak to peak; 2.5%.
3 60 pA.

Exercises 11

1 (a) 4.99 V; 3.6°; (b) 4.98 V; 5.7°; (c) 3.54 V; 45°; (d) 0.079 V; 89°.
2 (a) −0.26 dB; 35°; −3 dB; 90°; −5.4 dB; 108°; −12 dB; 137°.

(b) -1.2 dB; 49°; -4.7 dB; 90°; -6.3 dB; 105°; -13 dB; 131°;
(c) $+0.78$ dB; 35°; -0.39 dB; 90°; -3.1 dB; 113°; -11.3 dB; 145°.

3 $R = 796$ kΩ.
(a) $\frac{R_4}{R_3} = 0.59$, use say $R_3 = 10$ kΩ; $R_4 = 5.9$ kΩ;
(b) $\frac{R_4}{R_3} = 0.27$, use say $R_3 = 10$ kΩ; $R_4 = 2.7$ kΩ;
(c) $\frac{R_4}{R_3} = 0.96$, use say $R_3 = 10$ kΩ; $R_4 = 9.6$ kΩ.

4 (a) $f_1 = 195.1$ Hz; $f_2 = 205.1$ Hz; $\pm 45°$, $-45°$;
(b) $R = 563$ kΩ; $R_4 = 1$ kΩ; $R_3 = 4.93$ kΩ; 37 dB;
(c) $R = 398$ kΩ; $R_4 = 10$ kΩ; $R_3 = 19.5$ kΩ; 32 dB.

Exercises 13

3 $R_1 = 10$ kΩ; $R_2 = 15$ kΩ; $R = 118$ kΩ.

Exercises 14

2 $R_2 = 20$ kΩ; $R_1 = 10$ kΩ; $R = 50$ kΩ.

Exercises 18

2 1 mV/s.

Index